Data in astronomy

Data in astronomy

CARLOS JASCHEK

Centre de Données Stellaires, Observatoire de Strasbourg, France

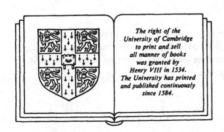

The right of the
University of Cambridge
to print and sell
all manner of books
was granted by
Henry VIII in 1534.
The University has printed
and published continuously
since 1584.

CAMBRIDGE UNIVERSITY PRESS

Cambridge

New York   New Rochelle   Melbourne   Sydney

CAMBRIDGE UNIVERSITY PRESS
Cambridge, New York, Melbourne, Madrid, Cape Town, Singapore,
São Paulo, Delhi, Dubai, Tokyo, Mexico City

Cambridge University Press
The Edinburgh Building, Cambridge CB2 8RU, UK

Published in the United States of America by Cambridge University Press, New York

www.cambridge.org
Information on this title: www.cambridge.org/9780521177283

© Cambridge University Press 1989

First published 1989
First paperback edition 2011

A catalogue record for this publication is available from the British Library

Library of Congress Cataloguing in Publication data
Jaschek, Carlos.
Data in astronomy/Carlos Jaschek.
    p. cm.
Includes bibliographies and index.
ISBN 0 521 34094 2
1. Astronomy—Data processing. 2. Astronomy—Observations.
3. Astronomy—Documentation. I. Title.
QB51.3.E43J37 1989
5201.28'5—do9   88-9504 CIP

ISBN 978-0-521-34094-6 Hardback
ISBN 978-0-521-17728-3 Paperback

Contents

Preface   *page* ix
Acknowledgements   xi
Abbreviations for bibliographic references   xiii
1   Observations   1
1.1   The definition of 'observation'   1
1.2   Planning observations   5
      Notes on Chapter 1   8
      References   9
2   Observatories   10
2.1   History   10
2.2   National and multinational facilities   15
2.3   The 'new' observatories   19
      Notes on Chapter 2   22
      References   23
3   Data   24
3.1   The nature of data   24
3.2   Models and theories   26
3.3   Data in teaching   27
3.4   Types of data   27
3.5   Presentation of data   29
3.6   Units, errors and standard values   29
      Notes on Chapter 3   33
      References   34
4   Archiving of observations   35
4.1   History   35
4.2   Support media   37
4.3   Data archives   38
4.4   Why do we preserve data?   40
4.5   What should be preserved?   40
4.6   Types of data archives   42
      Notes on Chapter 4   47
      References   48

5	**Presentation of astronomical data**	50
5.1	Data in research papers	50
5.2	Catalogues	52
5.3	Errors in printed catalogues	55
5.4	Computer-readable catalogues	56
	Notes on Chapter 5	57
	References	57
6	**Designation of astronomical objects**	58
6.1	Designation of stars	58
6.2	Designation of objects in general	61
6.3	Dictionary of synonyms	62
6.4	Designations and positions	62
6.5	Officially accepted designation practices	65
6.6	Designations and priority rights	69
	Notes on Chapter 6	71
	References	71
7	**Catalogues**	74
7.1	Surveys of catalogues	74
7.2	Data on stars	75
7.3	Non-stellar objects – general	80
7.4	Non-stellar objects in our Galaxy	80
7.5	Non-stellar objects outside our Galaxy	81
7.6	Availability of catalogues	83
7.7	Deficiencies of the present coverage	83
7.8	Inadequately-covered areas	85
	Notes on Chapter 7	86
	References	87
8	**The growth of data**	93
8.1	Growth characteristics	93
8.2	The growth of astronomical information	94
8.3	Incompleteness of our knowledge of the Universe	104
	Notes on Chapter 8	107
	References	108
9	**Data banks and data bases**	110
9.1	Storage of information	110
9.2	The establishment of a data bank	112
9.3	Requirements introduced by the use of computers	118
	Notes on Chapter 9	123
	References	124
10	**Data centres**	126
10.1	The establishment of data centres	126
10.2	Organizational aspects of a data centre	128

10.3   Integrated data bases   133
10.4   The future   141
       Notes on Chapter 10   141
       References   142
  11   **The publication of scientific information**   144
11.1   Historical development   144
11.2   Forms of publication   155
11.3   The publication of data   156
11.4   How to search for relevant literature   158
       Notes on Chapter 11   160
       References   160
  12   **The growth of scientific information**   162
12.1   The relative importance of different types of publications   162
12.2   How much of the information is useful?   166
12.3   The relative impact of journals: 'Bradford's law'   172
12.4   Publication and productivity: 'Lotka's law'   174
12.5   Evaluation of scientific activity   177
       Notes on Chapter 12   180
       References   182
  13   **International data organizations**   183
13.1   The International Council of Scientific Unions (ICSU)   183
13.2   The International Astronomical Union (IAU)   184
13.3   The International Union of Geodesy and Geophysics (IUGG)   189
13.4   The Federation of Astronomical and Geophysical Services (FAGS)   190
13.5   Committee on Data for Science and Technology (CODATA)   190
13.6   Committee for Space Research (COSPAR)   192
13.7   International Council for Scientific and Technical Information
       (ICSTI)   192
13.8   Conclusion   193
       Notes on Chapter 13   194
       Index   195

Preface

Natural sciences are based upon observations and experiments and the outcome of both is data. To be usable, data must be preserved on some support medium, such as paper. These two sentences summarize the objective of this book: a survey of how data are obtained, compiled and published.

Readers will find in Chapter 1 a discussion of observations. Chapter 2 deals with the places where these observations are carried out, the observatories. Chapter 3 deals with data in general. In Chapter 4 the ways observations are archived are discussed. Chapter 5 discusses the presentation of astronomical data for further use. Chapter 6 deals with the problem of how astronomical objects are designated. Then, in Chapter 7, I present the catalogues, in different fields of astronomy, that provide the most readily usable data. The growth of the quantity of data with time is the subject of Chapter 8. Chapter 9 deals with data banks and data bases; data centres and networks are the subject of Chapter 10. Chapter 11 deals with scientific information in general and the ways this information is published. In Chapter 12 the growth of scientific information in general and some of the uses made of publications for evaluation are discussed. Finally, the thirteenth and last chapter describes international organiz ations dealing with data.

Throughout the book the word 'astronomy' is used in a restricted ser which excludes solar system objects. The choice of this restrictic explained in Chapter 1; it depends essentially on the fact that all c vations beyond the solar system are exclusively based upon the stud radiation received. They are thus of a passive kind, beyond the experimentation either today or in the near future. For instance send a spacecraft to Mars and make observations *in situ*. It is to that in the next decades such landings (or close-up visits) can t nearly all bodies of the solar system, and this creates, essentially, two astronomies with different methodologies.

Because of the rapid growth in the number of astronomers and of observing facilities, we have to cope with rapidly increasing amounts of new data and numbers of publications. It is essential for our survival as scientists to find ways to order this information, if only to avoid being buried under tons of paper. The organizations for dealing with data and information exist already but new techniques have forced us to adopt new solutions; however, not all astronomers may be aware of these new solutions. Therefore, it seems timely to provide an introduction to this fascinating subject. Because of its rapid evolution, I have preferred to write a broad outline rather than an encyclopaedia.

My involvement with data started a long time ago, at La Plata, when one of my students, H. Conde, proposed the idea of publishing as a catalogue a file that we were compiling for private use. This was in the early 1960s and, since then, I have been associated with the handling of data. This link became even stronger when I was appointed Director of the Strasbourg Stellar Data Centre (CDS) in 1975. This centre, which will be mentioned many times in the book, has been in many ways a privileged 'watch-tower' over what has happened in the field during the following years.

So, it is clear that this work is the outcome of many exchanges with my colleagues over the years, especially the staff of the CDS and the members of its Scientific Council.

Acknowledgements

I want to thank Mr C. Schohn for the illustrations and Mrs Boisselier for typing the manuscript.

I would also like to thank the following for permission to reproduce the figures listed:
– Columbia University Press for Figure 11.1;
– The *Publications of the Astronomical Society of the Pacific* for Figure 12.2.

I would like to express my special gratitude to my wife, M. Jaschek, and to Messrs J. Delhaye, F. Ochsenbein and E. Davoust, who have read the draft of this book, or parts of it, and made many comments. I hope that I have used their comments in the right way. I am very grateful to Dr Jacqueline Mitton who corrected not only the English but also a number of errors and mistakes and helped to improve the presentation. Finally I want also to thank all those who have contributed with the many 'private communications' dispersed through the text.

Abbreviations for bibliographic references

In the bibliographic references at the end of each chapter I have used a number of abbreviations for the titles of journals, which are as follows:

AA = Astronomy and Astrophysics
AA Suppl. = Astronomy and Astrophysics Supplement
AJ = Astronomical Journal
AN = Astronomische Nachrichten
Ap. J. = Astrophysical Journal
Ap. J. Suppl. = Astrophysical Journal Supplement
ASpS = Astrophysics and Space Science
BICDS = Bulletin d'Information du Centre de Données Stellaires
Mem RAS = Memoirs of the Royal Astronomical Society
MN = Monthly Notices of the Royal Astronomical Society
PASP = Publications of the Astronomical Society of the Pacific
QJRAS = Quarterly Journal of the Royal Astronomical Society

'CDS catalogue x' designates a catalogue not otherwise published. It can be obtained from all centres listed in Chapter 10.

1

Observations

1.1   The definition of 'observation'

Astronomy is a natural science and as such is based upon observation. If one were to try to define the term 'observation', one would be led to say that an observation is the detection of a signal, carried out by somebody at a given place and a given time, with a particular instrument and for a particular purpose.

Such a definition is rather broad and could also apply, for instance, to the measurement of one's own body temperature. To use the term in a strictly astronomical sense, one must qualify the nature of the signal detected: it must be either electromagnetic radiation or a high energy particle from an extraterrestrial object. This applies to all astronomical observations, except the sampling of lunar soil and meteorites that have landed on the Earth, which can be analysed in laboratories. The only contact the astronomer has with the object of study is, then, through an electromagnetic signal or high-energy particles from an extraterrestrial object.

Astronomical observations differ from observations in physics and chemistry because, in astronomy, the place of observation and the date must be carefully specified. In chemistry this is unimportant in general.

We have still to add the very general requirement that an observation is only usable if it is documented. However, this applies to all sciences, not only to astronomy.

Let us next illustrate our definition with some examples. We start with Hipparchos (*c.* 150 BC), who observed the simultaneous rising and setting of pairs of stars on the sky. He observed visually, helped with a celestial globe, on which he marked the positions of his pairs of stars. He described his observations in *Comment on the Phenomena of Aratos and Eudoxus*.

Chinese annals describe the visibility of many solar eclipses at the times of kings of different dynasties. The observations were carried out visually by astronomers and a description of the cities where they were made was given in the Annals. These observations constitute, even today, important checks for the theory of the Earth's motion.

The fall of a shooting star is an observation if we are told when and where it was seen, which part of the sky was transversed and how long the phenomenon lasted.

A photographic plate taken with a camera having an objective prism provides the spectra of several hundreds or thousands of stars. If we are told which was the instrument used, the objective prism angle, the photographic emulsion, the dispersion of the prism, the place and date of observation and the exposure time, we consider this to be an astronomical observation.

If a satellite measures the X-ray flux emitted by a source in the sky, it only qualifies as an 'observation' if a (very) large number of parameters is specified – the position of the satellite, the description and specification of the detector used, the signal measured, the time and duration and the location of the source in the sky.

All the preceding examples insist on certain aspects of the definition given already. Let us examine now in some detail the different elements included in the definition.[1]

(a) Detection of an electromagnetic signal or of high energy radiation. We have remarked already that this becomes a rule for all astronomy of objects beyond the Moon. If we consider that, perhaps in the next decades, we will get soil samples from other bodies in the solar system, we may generalize and say that this condition will describe essentially all 'non-solar-system' astronomy. From here onwards, we shall use the word 'astronomy' in the sense of being 'non-solar-system' astronomy. A segregation of solar system astronomy from the rest seems reasonable in the 1980s when the methods used differ increasingly, the literature on solar system bodies has grown enormously and astronomers are increasingly specialized in either one or other of the two fields.

(b) Made for a particular purpose. An underlying purpose distinguishes making an observation from casually 'looking at the sky', as we all do from time to time. It is one of the most important factors that needs to be specified in order to make the observation properly. The purpose determines the technique used for carrying out the observation and puts it in its context. For instance, if somebody determines the geographical latitude by means of observations, one should know if the observations were carried out for land-surveying purposes, or for determining the position of an observatory, or for studies of the polar motion. If one does not know the purpose but only the resulting latitude, one may run into trouble: was it an approximate result obtained by a surveyor, expressed to more significant figures than were actually measured? Was it an eighteenth-century determination carried out by an astronomer?

Notice also that observations can be made for a certain purpose and used later for a completely different one. Messier (1781), for instance, was interested in comets and his list of objects was set up to avoid confusion with comets. For us it is the first long list of non-stellar objects. Similarly, Flamsteed measured the positions of stars and included also some observations of a planet, not recognized as such by him; these are the earliest observations of Uranus.

Supernovae and novae, called 'guest stars', were observed by Chinese, Korean and Japanese astronomers to predict the destinies of their kings, but today they are irreplaceable records for the study of supernovae and novae.

(c) Made by somebody. This is an important element of the observation, which goes with the documentation. It constitutes a recognition of intellectual property, like, for instance, the name of the discoverer of a comet or a variable star. By providing a name, an indication is given of the approximate epoch and, perhaps, even of the place at which the observations were made. The mention of the epoch also furnishes an indication of the possible techniques used. If other observations by the same scientist are known, it provides in addition a 'quality mark'.

Difficulties with authorship occur frequently and I quote a few. First of all, if the observations are of the routine type, they may have been carried out not by the astronomer himself but by one of his assistants and/or students. If a discovery results, who is to be regarded as its finder? (This happened, for instance, with the discovery of pulsars.) The observation may be made by several people, like in the case when somebody takes a plate, somebody else examines it and finds a suspected object and a third person measures it and discovers that it is a lost comet. Who is the 'observer'? Or the observation may be carried out automatically (see Chapter 2); the observation may then be associated with the astronomer who proposed that it be made.

It is clear that the usefulness of knowing the name of the observer diminishes with the passing of time. Except historians, most astronomers would be satisfied by knowing that they are dealing with a 'fifteenth-century observation'. Thus, observations are to be regarded like the stone blocks of the cathedral of Strasbourg – each one is important and is the personal work of somebody but, once put into its place in the building, it loses its individuality.

(d) Carried out at a given place and at a given time. The recording of these is a fundamental requirement for any observation and their absence practically eliminates the observation from further use in a science in which everything is variable with time. Although astrometrists are usually specific on these details, astrophysicists have become 'sinners' and it is seen nowadays, in quite fashionable journals, that 'photoelectric observations in the uvby system were carried out (with a precision of a few milli-magnitudes) in 1987'. Some years later, nobody is able to provide the exact dates of observations because the records were destroyed, the author moved to another place or is engaged in other research. The observations are thus useless, for example, for investigating time variations. Authors reply to such criticism that journals do not want observing dates (because it takes space) but they forget that, for instance, data banks are there for just such purposes. (This will be discussed in Chapters 9 and 10.) Observing places are also often omitted under the assumption that they say little about the observations; what is forgotten is that they are often the last hope for anybody looking for more details – an oldtimer of the observatory may still be able to provide details on the reduction of data one is interested in, or log books may be found in some drawer.

(e) Made with a technique that has to be described in detail. This is a self-evident condition for observations to be useful. If we are told that something was observed but not how it was done, we are in the positions of historians who try to deduce or guess from data how they were obtained.

It goes without saying that the techniques should be described in detail, keeping in mind that observations are preserved for future users and, sometimes, for uses other than the original ones. A few examples may be appropriate, although we shall consider this in more detail in Chapter 3.

Photoelectric observers from 1920 to 1950, for instance, described their photometers in detail but usually did not provide the average wavelength of the photocell–filter combination; in other cases they did not use filters at all. The result is frustrating: we have the observation of a flux with 1% precision but we are ignorant of the wavelength at which it was measured! Unhappily, most photoelectric photometry done before 1950 was of this kind.

Other observers describe their photographic material as Panchromatic Super X, but forget to say to what wavelength range that corresponds. If the manufacturer of the material is not indicated, the lack of wavelength specification may prevent further use of the observations. Similar remarks apply to descriptions like 'the plates were taken with the Mt. Apple telescope at Hollywood with the 10-inch objective prism'. It would eliminate a lot of literature searching if the plate dispersion and the wavelength range of the spectra had been indicated.

This rapid overview shows that complete documentation is required to make an observation fully usable. This is a seemingly obvious point and I think that we are all convinced that we have satisfied it. What we often do not realize is that many details that seem obvious to us *now* (so that it is unnecessary for them to be written down) will not be obvious at all for next generation of astronomers. It is just the non-mention of the obvious that complicates things.[2]

From the preceding discussion it is clear that an observation must fulfil a number of conditions to be usable. Not all of these conditions are equally important in all cases but if some are absent the observation loses some value and may become partially or totally useless. Since all astronomy is based upon observations, this means that we are 'building on sand'.

1.2   Planning observations

We turn now to the practical aspects of observing. We start by noticing that observations are usually carried out at places called observatories; we shall deal with them in Chapter 2. Secondly, we notice that one rarely makes just one observation; usually, a whole series of observations of similar type are carried out. This is called an 'observation programme'. Because any observation requires instruments (e.g. a telescope), auxiliary equipment (e.g. a spectrograph), assistance from other persons (e.g. an engineer or a night assistant), transport to and from the observatory, lodging after the night is over etc., it is clear that an observation programme must be planned very carefully in advance. For ground-based observations this means typically six months beforehand, whereas, for space experiments, the planning period may extend to six *years*. Excepted from such provision are those happy (few) astronomers who have an instrument available for their own use, which may happen when one uses an instrument not much needed otherwise or the prototype of a new instrument, or one is the owner of the telescope. For the remaining 98% or so of astronomers, observations must be carefully planned long in advance.

This implies that one must plan what to study, how and when; in other words, one must plan one's research well in advance of making the observations one needs. Let us add that careful planning does not guarantee that one will be able to carry out the programme because, in a similar way, many colleagues also plan to observe with the same telescope. In the ensuing competition, the best proposer usually wins; we shall see in the next chapter how the competition is decided in practice by the 'programme committee'. It is clear that one increases the chance of succeeding if the purpose of the observing programme is well-founded (i.e. it is clear why one wants to observe), if the proposal contains a reasonable request for observing a limited number of objects (since, if the programme is too extended, it will collide with more programmes than if it is short) and if one provides a clear explanation of why one needs a specific combination of telescope and auxiliary equipment.[3]

What kinds of programmes can be proposed? In general, there are two extreme types: the survey type and the individual, monographic type. We call 'survey type' all those programmes in which the emphasis is put on the observation of a large number of objects with the same technique, and 'monographic type' the one in which the concern is for one (or a few) objects to be studied in detail. Because of the number of objectives involved, it is clear that one might also speak of long-term programmes and of short programmes – although a long series of monographic-type programmes may constitute a future long-term programme. The terms 'long' and 'short' refer in this context more to what one writes in the proposal asking for observing time than to the long-term philosophy. It is clear that the two types of programmes have different objectives. In a survey programme one is more interested in the statistical aspects of the objects, such as distribution in space, average properties, correlation between parameters and so on. To know the distribution of B-type stars on the sky, or that of galaxies in space, for example, it is clear that one must study as many objects as possible, even when one does this only for a part of the sky. It is also the only way of picking out interesting objects, or to find out what is a 'typical' (i.e. average) object of the type one is studying.

The drawback of survey programmes is that they demand much time for their execution. Since telescope time is often at a premium, programme committees tend not to be overenthusiastic with such proposals. This is often justified on the ground that surveys tend to produce fewer papers per year than monographic studies. We shall discuss in Chapter 12 how to evaluate publications; here we observe simply that the impact of surveys is often much greater than that of monographic papers. In favour of monographic studies one can say that the newest instrumentation may be used to study one (or a few) parameters of one object. One may thus expect real progress or new insights, which is the best reason for such studies. One examines, you might say a small section of the universe with a magnifying glass. To safeguard against excessive optimism, let us formulate two reservations. The first concerns the selection of the object: one should make sure by all conceivable means that one is dealing with an average (typical) object; otherwise we might learn much about an interesting object but not very much about the whole class of objects. (Consider, for instance, if extraterrestrials decided to study mankind and considered as 'average' either a man named Johann Sebastian Bach or another called Gengis Khan.) The second reservation concerns the use of new instrumentation: usually one underestimates the investment of time that goes into the making of a new instrument and getting it to work properly. For instance, in space astronomy, the time for development may well be of the order of five years or more.

We have already said that monographic studies demand less observing time (which makes them popular with programme committees) and generally produce more papers per unit of observing time than survey programmes. We have emphasized the differences between the types of programme one may carry out, because the choice may well influence the scientist's life for many years. It is clear that a careful choice must be made.

Usually, the young astronomer is not concerned with such choices because thesis work is made under the supervision of a senior scientist who directs the student. Since postgraduate work is usually connected with thesis work, the necessity of choice may come up only years later, when, for instance, the astronomer wants (or has), to re-orient his research.

We may add two more considerations to 'planning'. First of all, one must also consider the available observational facilities.

The variety of instruments and equipment available to an astronomer in an industrialized country should not obscure the fact that such good fortune is not shared by all astronomers in all countries, nor at all times. Western Europe, for instance, had no modern telescopes until the 1950s and was therefore rather backward in observational astrophysics; Latin American astronomers today do not all have access to modern instruments. It is important to consider such limitations from the start. The astronomy one may do in one's lifetime often does not depend so much on one's own wishes as on what the society that support us provides us with.

A second consideration one should always keep in mind in the planning stage is the amount of work it takes to obtain the data from the observations, i.e. the data reduction. This is often underestimated, with ensuing difficulties. The history of astronomy is full of projects that were too large for the time allowed and that were afterwards either abandoned or not fully carried out. It seems that each country has had its own 'white elephant'! The analysis of such cases is often very illuminating for the development of science, but falls (regrettably) outside the scope of the book.[4]

Before closing the chapter, we should mention a last point, which arises in monographic studies and is due to the technical progress that opened to astronomers a number of wavelength regions other than the visible one (350–670 nm). In succession, the radio, ultraviolet, infrared, X-ray and γ-ray regions became accessible, and so it became interesting and fruitful to study an object in different wavelength bands simultaneously so that time variations do not affect the intercomparison.[5]

The problem of how to arrange simultaneous observations from different observatories has no easy solution. A colloquium at Strasbourg, *Coordination of observational projects*, in late 1987, will deal with these problems and the reader is invited to consult the proceedings.[6]

One obvious conclusion is that a careful planning of observations well in advance is needed more than ever.

Notes on Chapter 1

1. Curiously, it is hard to find in the astronomical literature a definition of the term 'observation'. For observation and experiments in sciences see Ziman (1978). For a philosophical point of view, concerned specially with the observer, see Kutschmann (1986).

2. One of the first astronomers to call attention to the need for documenting observations carefully was Lacaille (1713–62) who wrote that . . . 'in the advanced state of astronomy, no-one could any longer be believed on his mere word', and that, in order to employ with confidence an observed position, it was necessary to have all details of the observation and all the elements of reduction.

3. Just to illustrate how important a programme proposal is nowadays, we may quote Heck and Egret (1987), who say, 'in fact, scientists are now complaining that writing a good observing proposal requires as much time, care and energy as a paper for a refereed journal'.

4. An example of a good programme that ran into difficulties when it came to the reduction of the material, is the *Carte du Ciel–Astrographic Catalogue* project. Started in the late 1880s, it was to photograph the whole sky, with similar instruments installed at different latitudes. A short exposure series was to be measured and reduced immediately – the so-called *Astrographic Catalogue* –

whereas the long-exposure series, the *Carte du Ciel*, was to be considered as a document usable by future generation of astronomers. This generous project, proposed by French astronomers, was a success with respect to observing, but turned into a painfully lengthy operation when it came to reductions. All measurements had to be done by eye, using hand-moved screws or rasters, and subsequent plate reduction had to be made by hand or by mechanical calculators driven manually. No wonder that it took over sixty years to complete the *Astrographic Catalogue* and that, for the two countries which were most involved with it (France and Australia), it turned out to be a crushing burden that left little time for other work. In retrospect, this was clearly due to faulty planning – the amount of reduction work was considerably underestimated. The reader may find more details in the proceedings of IAU Symposium No. 133, *Mapping the sky: past heritage and future directions*, which took place in June 1987 at Paris (Debarbat *et al.*, 1988).

5. Another reason for carrying out coordinated observations is the need to cover entire cycles of a given variable star. This clearly needs observations from places located at different geographic longitudes.

6. Another meeting on the same subject was called *Multiwavelength astrophysics*. Proceedings of both are to be published by Cambridge University Press.

References

Debarbat, S., Eddy, J. A., Eichhorn, H. K. and Upgren, A. R. (ed.) (1988) *IAU Symp. 133, Mapping the sky: past heritage and future directions*, Reidel

Heck, A. and Egret, D. (1987) *The Messenger* **48**, 22

Hipparchos, *Comments on the phenomena of Aratos and Eudoxus*, translated by C. Manitius, Leipzig 1913, B. G. Teubner Verlag

Jaschek, C., Hernandez, E., Sierra, A. and Gerhardt, A. (1973) *Catalogue of stars observed photoelectrically*, Publ. Ser. Astr. La Plata, **38**

Kutschmann, W. (1986) *Der Naturwissenschaftler und sein Koerper*, Suhrkamp Verlag

Messier, C. (1781) *Connoissance des temps pour l'année bissextile 1784*, Paris, p. 227

Ziman, J. (1978) *Reliable knowledge: An exploration of the grounds for belief in science*, Cambridge University Press

2

Observatories

2.1 History

In Chapter 1 we used the term 'observatory' in the sense 'place where observations are made' without being specific. In this chapter we shall examine briefly the evolution of observing places over history and the current system for obtaining observations. This is an essential step for understanding how one makes the transition from observations to data. However, the reader should remember that this is not a book on the history of astronomy, so only a very sketchy outline of the subject is given.

The first groups to show systematical interest in phenomena happening in the sky were the priests of primitive societies. They observed the Sun, the Moon, the stars, the planets and also a wide range of meteorological phenomena like clouds, coloured sunrises or sunsets, lunar halos and storms. We can only speculate on their motivation: perhaps they were mostly interested in discovering the will of their gods as manifested through the phenomena observed.

We have, for instance, clay tablets from astronomers, priests of Uruk, 1650 BC, providing observed positions of Venus. Such reports extending over decades denote that an organization of observers and reporters of heavenly phenomena existed. Since the reports were made by priests and addressed to kings, it can be surmized that the observations were made from the ziggurats or the palaces. We know what the motives of these astronomers were: they observed to know and predict certain solar-system phenomena.

When Greek visitors brought this knowledge to Greece, the motivation changed. Since the Greek visitors were 'philosophers' (i.e. intellectuals) and not priests, they tried to understand nature. Thus, the search for celestial omens was neglected for some centuries. Greek astronomers observed at their own cost and, since no large instruments existed, observing was probably done from the astronomer's house.

In the second century before Christ, the Kings of the Ptolemaic dynasty established at Alexandria a new type of institution, the 'Museum'. Participating scientists were protected and paid by the kings, who probably also paid for the instruments used for further observations and the scribes for copying the manuscripts for the library. In many ways, the Museum was the forerunner of the modern universities. We know the names of many great scientists associated with the Museum, including several astronomers. It was the last grand Alexandrine astronomer, Claudius Ptolamaeus, who codified most of what we know of ancient astronomy. In his writings he discusses the instruments he used but does not speak of an observatory, so that one may surmise that such a thing did not yet exist as an entity distinct from the Museum. In Hellenistic times, astronomy became strongly related to astrology, i.e. the art of predicting the destiny from the positions of the Sun, Moon, planets and some stars at the moment of birth. Astrologers were usually not observers, contenting themselves to use the ephemerides established by astronomers. This is, incidentally, the reason why astrology appeared only at a time when astronomy had progressed far enough to produce reliable predictions of the positions of the bodies of the solar system. The interest of people in astrology helped to keep the interest in astronomy alive at times when other sciences fared less well.

Partly for practical reasons (the calendar, the orientation of mosques toward Mecca) and partly because of astrology, astronomy flourished in the Muslim civilization. The first Arab caliphs began to attract scientists to their courts (including astronomers), ending up with scientific establishments that paralleled the 'Museum'. We know, for instance, that Caliph al Mammun (*c.* 830) constructed an observatory. An observatory can be defined as a place where systematic and continuous observations were carried out: it is a place with buildings to lodge the instruments, the library and the astronomers. We are poorly informed about the numbers of astronomers in such places but, probably, there were just a few – professors and students, paid by the royal treasury. We know of at least four Muslim observatories between the tenth and fourteenth centuries, located at Baghdad, Cairo, Samarkand and Maraga, but most of them had a short life because of the political instability. Most of the Arabic astronomical research was nevertheless still carried out by scientists who moved from one place to another and transported their instruments with them.

Better conditions for long series of observations existed in China. There one finds that there was a complete organization for the observation of the sky, the calendar and for the observation of omens, from the fourth century before Christ onwards. Thus, it seems clear that observatories were found mainly in countries having both a stable political organization and an interest in celestial omens.

When we pass to the Western world, we find at first isolated scholars, such as the monk Eginhard (*c.* 800), or groups of scientists like the one assembled by king Alphonse X of Castile at Toledo (*c.* 1260) to carry out certain astronomical computations and observations. In the Middle Ages there seems to have existed no regular observatory (Pedersen, 1976) in the modern sense. The first establishment of a permanent type that we know well was the Uraniborg Observatory on the island of Hven, built by Tycho Brahe around 1580 and financed by the Danish King, Christian II. The main task that Brahe had set for himself was to observe with the highest possible accuracy the positions of all celestial objects. Brahe described in detail both the observatory and its instruments, in which he took great pride. The instruments permitted measurements to an accuracy never obtained before, of about four minutes of arc. The observations were made by Brahe and a number of assistants, lodged at Hven.

The financial cost of the enterprise was accordingly high and, since Brahe himself was quite a character, it is no wonder that, after the king's death, the observatory was rapidly dismantled.

The results of Brahe's activity could be used immediately by his later associate Johannes Kepler to derive the laws of the motion of the planets, a fact which showed that progress in astronomy could be made through properly organized observations.

The next nations that built observatories had more solid treasuries to finance them on a long time scale – they were France, where Louis XIV founded the Paris Observatory in 1667, and England, where Charles II founded the Greenwich Observatory in 1675. In the Charter of the Greenwich Observatory it was clearly stated that the observatory was built . . . 'in order to find out the longitude of places for perfecting navigation and astronomy'.

The status of the observatories was thus due, at least in part, to the fact that astronomy had become a science with practical applications for navigation on the ocean and for establishing accurate maps. Such utilitarian purposes were, of course, handicaps for the purely scientific side of astronomy, but they helped establish an observatory in each of the bigger and stabler countries. This led to a proliferation of observatories during the next centuries, usually imitating either Greenwich or Paris. Fig. 2.1 shows the number of new observatories founded, for each decade, as a function of time. The growth seems to be exponential, at least for the nineteenth century, as shown by Hermann (1973), from which the data were taken. After 1900 the growth is declining. This is typical of logistic growth, which we shall consider in more detail in Chapter 8. Let us add that this rapid growth implied a similar increase in the number of astronomers during these times: nineteenth-century observatories were usually staffed by a few – certainly less than ten – astronomers.

This nineteenth-century exponential growth is due primarily to the creation of a number of new universities, which often had a chair of astronomy and some observing facilities. Observatories became identified with 'civilization' in a broad sense. As an example we may quote the discourse of the Argentine President Sarmiento at the dedication of the Córdoba Observatory: 'No nation may pretend to be a civilized one if it has no observatory'.

Before going into the further evolution of observatories, let us define the meaning of the term 'observatory', in the nineteenth-century sense. An observatory is a place where systematic observations of astronomical objects are made and research based upon observations is performed.

Fig. 2.1. Number of observatories as a function of time, from Hermann (1973). The analytical expression (open circles) represents $N = 12 \exp [0.0275 (t - 1800)]$.

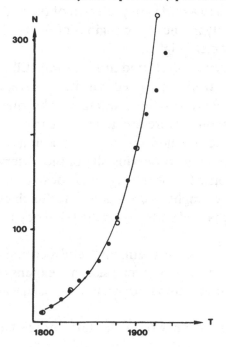

Materially, an observatory comprises buildings to lodge instruments, the library, the astronomers, the technicians and eventually some workshops (mechanical and, more rarely, optical).

Around 1850 a rapid growth in the application of new ideas in the fields of optics, photometry and spectroscopy to the study of astronomy started. This new branch was called astrophysics to distinguish it from astrometry, the measurement of the positions and motions of heavenly bodies, which is essentially what astronomy was up to the middle of the nineteenth century.

Toward the end of the century it had become clear that astrophysics needed relatively complex (and expensive) new equipment and increasingly larger (and more expensive) telescopes. Because of the availability of large private funds in the United States, American astrophysicists got increasingly bigger instruments. We find, in succession, telescopes of 152 cm aperture (1908), 252 cm (1918) and 508 cm (1949), all set up on the West Coast of the United States, whereas, in 1950, the largest European telescope was still one with a diameter of only 125 cm.

Most of the new American telescopes were installed on mountains: the Lick Observatory on Mt Hamilton and the Mt Wilson Observatory on Mt Wilson. The installation on mountains avoids the lower layers of the atmosphere, guaranteeing better seeing. Furthermore, these mountains were chosen to be far away from cities to avoid light pollution of the night sky and they were located preferentially in one of the regions of the world (California) with a high proportion of clear nights.

The Lick Observatory (whose construction started in 1875) was still an observatory in the old sense that all astronomers lived 'on site', whereas, at Mt Wilson, only the telescopes were on the mountain. The main facilities (offices, libraries and workshops), were erected in a nearby city (Pasadena) and the astronomers had no housing in the observatory grounds. The mountain station was thus the observing site of the observatory, at which astronomers stayed only for observing runs. Besides the astronomers, there was also a staff of night assistants and technicians, residing on the mountain, who were charged with having the telescope run smoothly.

Such an arrangement turned out to be very effective and became a model for many other institutes. It was, however, also an expensive arrangement because it split the old observatory into two bodies, each with its own needs.

Let me add that the idea of having one or two smaller telescopes of an institute at a better location than that of the main observatory is much older: Harvard Observatory had for many years a station at Arequipa (Peru), for instance. But what the Mt Wilson Observatory implied, was a new thing – it separated neatly the observing site from the site where the observations were analysed. It also suppressed the old idea of having the astronomer's living quarters on the observatory grounds. This dissolution of the old type of observatory is visible even in the architecture of the buildings – whereas the old style observatory often had buildings that were elaborate architecturally, the 'new generation' observatory buildings are very sober and functional.

2.2   National and multinational facilities

In Europe, the First World War cut short the plans for new observatories. It was only in 1936 that France started the Haute Provence Observatory. This institute was to be the national observing station for all French observatories – we shall call it 'national observatory' for short. Such a solution permits the pooling of the resources of many smaller institutes that are incapable of financing individual observing stations on their own. Interrupted by the Second World War, the Haute Provence Observatory was only really developed in the nineteen fifties. The idea of a 'national facility' was very soon adopted by other countries, such as the United States (National Optical Astronomy Observatory at Kitt Peak).

The last step in the evolution is represented by multinational observatories, where several countries develop one big observing facility, such as the European Southern Observatory in Chile, the telescope at Hawaii shared by French, Canadian and US astronomers and the Anglo-Australian Observatory. Obviously, such solutions became practicable only when air transportation permitted rapid travel, and various media (telex, radio) assured rapid communications with the home institutes. Such multinational solutions pool the resources of many countries, permitting the construction of telescopes that are much larger than those each country could afford separately. The prerequisite for such solutions is that very smooth relations exist between the host country (where the telescope is erected) and the member states that pool their resources.

I should remark that the national or multinational observatories were never planned to replace the existing observatories: they are simply the place 'where observations are carried out'. Each visiting astronomer carries back the observations to his home institute, where he converts them from raw or calibrated observations to analysed data (see Chapter 3 for details) and where he carries out his research. The functions (observing and research) of the old observatories are now redistributed, with observing being done at a national facility and research at the home institute. It is somewhat confusing to call both 'observatories'! Because of their different role, the staff of national facilities have as their main mission facilitating the observations of the visiting astronomers; their personal research comes only second, and they are not even allocated much more observing time than other applicants. On the other hand, the main job of the staff of the 'home institutes' is doing research, rather than observing (besides, of course, teaching, popularizing and administrative activities).

To be fair, it should be added that, in compensation for the work they are providing for the community, staff members of national facilities have certain advantages: knowing exactly the state of instrumentation available, which permits the best choice for their own research; benefiting from programmes given up (because of sickness for instance); awareness of the latest observations carried out for certain types of objects, and so on.

On the other side, the number of staff at a national facility should never be permitted to grow too much, because they would then constitute a powerful interest group whose interests could conceivably collide with that of the national community – in other words, they may wish to become a 'mega-observatory'. Such tendencies seem even more dangerous in multinational institutions.

Among the many problems that exist in the case of both national and multinational observatories is the one of the distribution of the observing time among the different participants. Whereas in the old observatories such a distribution was made either autocratically by the Director or more democratically by the staff, such solutions are impracticable when various institutions participate. Over the years, the idea has evolved of a 'programme committee', composed of astronomers chosen because of their expertise and their impartiality. This Committee receives and judges all proposals for observations ('observing programmes') formulated by the astronomers wishing to use the instruments. The proposals are judged only on scientific merit and the available time is distributed among the best proposals. This is a difficult task since the 'pressure factor' (defined as the ratio between the 'time asked for observation' over the 'time available'), is usually larger than one. Due consideration has also to be given to long-term programmes (of the survey type), to the observations of exceptional phenomena (bright novae or comets, for instance) and to the time needed for mechanical or optical work on the telescopes. Despite all these

difficulties, the procedure of having the time distributed through a programme committee seems superior to a distribution of time according to
other criteria like authority, nationality or 'equal chances to all applicants'.
Difficulties may appear if one of the partners uses financial considerations,
saying that if partner X has borne 28% of the costs it should also have (at
least) 28% of the observing time. Such constraints must be carefully
assessed by the 'programme committee' but sometimes they result in
observing time distributions that tend to favour the largest possible
number of applicants, rather than to concentrate on fewer, more promis-
ing, projects. Such discussions have led to open criticism of national
observatories (Harwit 1981). According to critics, national observatories
are less fruitful in making discoveries than observatories where telescopes
are only for staff members. National observatories have, on the other
hand, enabled many astronomers from small institutes to get access to
large observing facilities.

Abt (1985) has tried to assess the importance of the contributions of four
American institutes that have large optical telescopes. He chose two
national observatories and two university observatories. He then analysed
all publications derived from observations at the biggest telescope of each
institute; the analysis was done according to evaluation techniques
described in Chapter 12. He concludes that, if differences do exist, they are
not statistically significant.

On general grounds it must be recognized that national or multinational
observatories are the only practical solution to

  (a) access to large instruments for astronomers working at small
      institutions;
  (b) the erection of observatories at the best observing places (Hawaii,
      Chile, Canaries, South Africa);
  (c) the increasing costs of telescopes and, more general, of astronomi-
      cal research.

So, in the end, it seems that the advantages outweigh the disadvantages
quoted.

One should also give some consideration to the sizes of the observatories. We have seen that all old observatories were rather sparsely
staffed – be it only because of the problem of housing a large number of
astronomers on observatory grounds. Because of the increasing complexity of instrumentation and the rapid expansion of astrophysics, the size of

Table 2.1. *Staff size of the Royal Greenwich Observatory*

1675–1822	Astronomer Royal + 1 Assistant
1822–1835	Astronomer Royal + 6 Assistants
1835–1881	Astronomer Royal + Chief Assistant + 8 Assistants
1900	18
1930	42
1950	185
1980	234 (Part-timers counted as one-half post; includes La Palma staff)

Note: The table includes only established staff members; temporary staff are excluded. Sources provided by J. Dudley, librarian and archivist, RGO. For 1900, 1930, 1950 and 1980 the data were taken from the Annual Report.

the staff tended to increase over the years and there exist today many institutions that have a staff of more than a hundred.

I have assembled in Table 2.1 some figures showing the evolution of the staff numbers at the Royal Greenwich Observatory, which can be taken as a representative case. In fairness it should be added that the RGO now runs an important station at La Palma on the Canary Islands – the staff there is included in our counts, which explains the increase in the last line. Even with this consideration, the tendency is clearly of sustained growth. In view of this tendency, one can ask the seemingly odd question of whether such growth is producing better astronomy. Although I am unaware of specific studies on the subject and can therefore only rely upon personal experience and common sense, it seems that somewhere there must be a 'maximum efficiency'. It is clear that a place with 20 staff members constitutes a better scientific environment than a place with only 5, but what about the difference between 200 and 100? It cannot be taken for granted that large staffs imply more efficiency, although the point can be made that big projects imply big institutes, like the Space Telescope Science Institute. The presence of such institutes in a country probably depends on its wealth and population.

Leaving aside such big institutes, one might still ask whether it is more reasonable to have many small institutes or just a few big ones. (This question is very important in countries with a few hundred astronomers.) Although it is hard to formulate a general answer, it seems that it would be better not to concentrate all efforts in one institute. Institutes tend to develop a style of their own, both in research and in teaching, and the suppression of smaller institutes in favour of just one big institute puts an end to healthy variety. The question of maximum efficiency can probably be reformulated by saying that, except for institutes dealing with a specific big project, e.g. European Southern Observatory, Space Telescope Science Institute, observatories should have staffs of the order of 10^1 to 10^2 members, and no country should let astronomy become excessively concentrated.

2.3   The 'new' observatories

Let us examine next what happened to the role of observatories when new techniques, such as radio astronomy arrived.

Observing with his own antenna, Jansky detected in 1931–2 the first astronomical radio source to be discovered. Later on came the classical observatories that installed radio receivers, such as Leiden around 1945, and then we got the national or international radio facilities like the NRAO (National Radio Astronomy Observatory) of the US, or the IREM (Millimeter Emission Radio Institute) at Grenoble. The development thus runs essentially parallel to that of optical observatories.

Due to rapid advances in signal detection in radio astronomy, the cost of radio astronomy is rather high – not only because of the need for increasingly larger flux collectors (telescopes) but also because of the need for building updated auxiliary equipment. This has the natural consequence that many of the smaller and poorer countries never got really involved in radio astronomy.

A similar thing happened with space astronomy, which is so expensive that only the richest countries may get into it. To be clear, we use 'space astronomy' in a broad sense: it comprises observations from stratospheric balloons (flying at about 50 km), from rockets (reaching higher altitudes but remaining only for a few minutes outside the atmosphere), and from artificial satellites. Probably in the next decades we shall also have to include observations made from lunar observatories. In the early days of space research, the biggest problems were the technical ones, i.e. how to build spacecraft and make them work properly. Space organizations thus tended to focus on this aspect, whereas the scientific aspects of planning and designing instruments was mostly left to research teams at various universities or observatories.

As experience has shown, it takes a very long time to conceive an instrument, to have it approved, built, tested, flown and used. The timescale being of the order of ten years, astronomers (or teams of astronomers) involved in such projects want, quite naturally, to be the ones who first use the new results. Usually such teams are given a privileged status, granting them exclusive use of the new data over a short period of time (from six months to two years). After this period, the observations become public domain knowledge and may be used by any member of the astronomical community. (For other solutions see Vette 1982).)

After this, the original teams usually break up to get engaged in other projects. This is serious, because in that way the community loses readily available expertise. Therefore, in recent years, the original teams have not been disbanded but only reduced in number, so that they constitute a core of experts on the project.

Despite such an improvement, one must still be concerned about the fact that each space project is handled by one or more teams located at geographically different places. No assurance can be given that Observatory X, interested in X-ray astronomy in the seventies, will still be interested or able to do it, say, twenty years later. And, in such cases, it is not at all clear what will become of the experts, of the expertise and of the observations obtained and analysed. In fact, a clear tendency exists now for a 'space institute' to be set up in each country involved in space research, which has the functions of

(a) planning space research;
(b) observing or distributing satellite observations;
(c) reducing observations, i.e. developing algorithms to change raw data to calibrated data;
(d) developing software to exploit calibrated data;
(e) archiving observations.

('Distributing' in point (b) refers to situations where country X is not the sole owner but only a partner in a satellite programme.)

Since 'space institutes' are new, there exists as yet no universal agreement on all their functions but one sees in general that they have a role similar to that of the 'national facilities' discussed before, and the role of the staff in such establishments is quite similar.

Thus we see that developments in space astronomy are repeating in a general way those that took place in optical astronomy.

Because of the analogies, one may ask conversely if some of the advances made in space astronomy could not be adapted to ground-based astronomy.

One good example is the way observations are carried out. In space astronomy, the instruments usually fly high up in space and are operated remotely from the ground, through a series of commands given by the engineers and the astronomers. On the other hand, things are very different in ground-based astronomy. If an astronomer is granted observing time at a national or multinational facility, the observer flies to the site (usually involving also a trip by car, an overnight stay to adapt himself to the change in altitude or longitude) and then observes at the telescope assigned to him for a (short) period. It is a fact of life that at big optical telescopes one usually gets runs of 2–5 nights, and not more (but often less) than one month per year of total observing time. In the case of an observer from the northern hemisphere observing in Chile, five days observing time implies about five additional days lost in travel and adaptation, so that one may ask (especially if not all nights were usable – something that may happen even in Chile!) if there are not better ways to organize observations.

A solution is remote operation (Longair *et al.* 1986). Several cases are conceivable:

(a) remote operation. The instrument operates under remote control like a satellite and no night assistant at the telescope is required. The observer monitors the weather and seeing conditions, manipulates the instrument and obtains observations transmitted to him.

(b) remote observing. Here there is a technical team working at the telescope (night assistants, engineers) but the astronomer is not present. He is told in real time about the output and can decide if things are going well and what should be observed.

(c) remote eavesdropping. Here the observers (not present at the telescope) is informed of what happens, but cannot intervene.

(d) service observing. The observations are carried out by the technical staff on site. The (absent) astronomer only provides the programme of what should be observed and how.

In this list, the traditional observation method, when the astronomer carries out the observations himself (assisted by the technical staff), would be case (e). It is clear that solution (a) is technologically the most complicated one, whereas (d) requires practically no new technology. The reader is referred to Longair *et al.* (1986) for the advantages and disadvantages of each solution but, in summary, one can say that astronomy is moving rapidly away from solution (e) toward solution (a).

The general dissociation of the place where the observation is carried out from that where astronomical research is done, has forced a change in the meaning of what an 'observatory' is: today one may define it simply as 'a place where astronomical research is carried out'. It would be more appropriate to talk of astronomy departments or of astronomical institutes in such cases and reserve the term 'observatory' for those places where observations are carried out.

For completeness we should remark that, even before the present time, observatories existed that had no telescope at all associated with them. Examples are the Astronomisches Rechen-Institut at Heidelberg, where the emphasis is on calculations, and the Bureau des Longitudes at Paris, devoted mainly to celestial mechanics.

The changing definition of the term 'observatory' illustrates very well the long history of astronomy from ancient times to the present.

Notes on Chapter 2

The evolution of observatories through history must be studied in books on the history of astronomy, for instance, Pannekoek (1961), because a specific book on the subject is lacking. The only book on the subject is from a mostly architectural viewpoint (Muller 1975). The book by Harwit (1981) contains many interesting discussions relating to this chapter. Gingerich (1984) contains several sections on the subject matter of this chapter, among them a description of several observatories, namely Greenwich, Paris, Pulkovo, Harvard College, United States Naval Observatory, Lick and Potsdam and the southern hemisphere observatories. The last paper – like several others in the series – is slightly unbalanced because of a rather Anglo-Saxon point of view. It is clearly difficult to write a completely neutral history!

For the early history of the foundation of American observatories, see Musto (1967). A list of the existing active observatories in the world is given by Heck and Manfroid (1987).

For the role of programme committees see the papers in *Coordination of observational projects in astronomy*, C. Jaschek and C. Sterken (ed.) (1988) Cambridge University Press.

As a final comment, it is striking that in the flood of several hundred books on astronomy per year (see Chapter 12), one finds none on the organizational aspects of astronomy. We have already remarked that no book on the role of observatories exists and the same is true of the internal organization (role of directors, boards, committees) or the organization of research (new telescopes, programmes). The material exists, of course, but there seems to be no public interest in such matters. It would also be most interesting to compare the evolution of observatories with that of other scientific institutes (e.g. physics or chemistry laboratories) of the same epoch. Astronomers tend to see their activities as something very specific – unique – and the awareness that astronomy is but a (small) part of science seems to be generally missing. A comparison of what an industrialized country spends on the different sciences shows that astronomy represents only a small slice of the total science budget and the same is true in terms of manpower.

References

Abt, H. A. (1985) *PASP* **97**, 1050

Gingerich, O. (ed.), (1984) *The general history of astronomy, vol. 4, part A (Astrophysics and twentieth-century astronomy)*, Cambridge University Press

Harwit, M. (1981) *Cosmic discovery: The search, scope and heritage of astronomy*, Basic Books

Heck, A. and Manfroid, J. (1987) *International directory of professional astronomical institutions*, Publ. Spec. CDS No. 9

Hermann, D. B. (1973) *Die Sterne* **49**, 48

Longair, M. S., Stewart, J. M. and Williams, P. M. (1986) *QJRAS* **27**, 153

Muller, P. (1975) *Sternwarten: Architektur and Geschichte der astronomischen Observatorien*, Peter Lang

Musto, D. F. (1967) *Vistas in astronomy* **9**, 87

Pannekoek, A. (1961) *A history of astronomy*, Allen and Unwin

Pedersen, O. (1976) *Vistas in astronomy* **20**, 17

Vette, J. I. (1982) In *IAU Coll. 64, Automated data retrieval in astronomy*, Jaschek, C. and Heintz, W. (ed.), D. Reidel, p. 243.

3

Data

3.1   The nature of data

We said at the beginning of the book that in natural sciences knowledge is based upon observations. But observations alone are only the raw material, which needs to be interpreted with the help of a model or a theory. The distillation of observations produces data, which are important cornerstones in all sciences.

We shall define data as parameters that can be deduced from observations by a convenient model or theory. Let us try to provide some examples, based upon the examples given for 'observations' in Chapter 1.

We mentioned Hipparchos, who described the position of stars on the night sky. By plotting the positions of the stars on a celestial globe – probably a metal sphere – he found that the stars had changed their positions relative to the vernal equinox, in a certain systematic way with regard to earlier observations of the same kind. Thus he discovered the phenomenon of precession due to a slow 'rolling' of the Earth's rotational axis. By comparing the existing observations, he deduced that celestial latitudes did not change, but that celestial longitudes increased by about 1° in 100 years. This value of 1°/100 years = 36″/year is the precessional constant, the numerical datum that summarizes his discovery. (The modern value is 50″.27.)

Then we considered the observation of solar eclipses by Chinese astronomers who wrote down where and when they were visible. A modern astronomer can extrapolate backwards in time the motion of the Moon and the Earth and verify whether these eclipses occurred at the dates we predict nowadays. If this is not so, we may introduce a correction term in the motion of the Moon, the so-called 'secular acceleration', and the analysis of the discrepancies provides us with the value of this acceleration, which is again a datum.

Then we considered the spectra of stars photographed with an objective prism. Each star has a position on the plate, which allows us to deduce its rough position on the sky. Each of these positions is a datum, giving the stars' coordinates. Furthermore, each of the objective prism spectra can be classified according to its similarity to a set of standard stars. The spectral types (adjudicated by visual comparison with spectra of standard stars) are also data, not of a numerical type, but shorthand descriptions like A2 or A2 V. We may study the spectrum of each star individually on the basis of calibrated spectrograms. In that case one may derive for a star its radial velocity, its surface temperature and its chemical composition: each one is a separate datum but all are derived from the same observational material.

When one compares data in astronomy with those in other natural sciences like physics, one difference turns up quickly. Let us take the velocity of light. When it was first determined by Roemer in 1675 from a study of the occultation of Jupiter's satellites, he got a value of $c = 227\,000$ km/s.

This value was improved over the course of time to 299 990, 299 802 and more recently to $c = 299\,792.5 \pm 1.0$ km/s but there is no hope that this is the final value, because the precision can still be increased. So, whenever a new value for the datum c arrives, the physicist is happy to drop the previous value, and adopt the last one. (Physicists finally adopted 299 792 458 m/s by convention.) Similar changes happen with all data in physics: the charge of the electron, the mass of an isotope, the gravitational constant.

An astronomer, however, cannot treat data in the same way because almost all astronomical phenomena are variable with time. Each observation leads to a datum, which is the only one available to us in the years to come. For instance, we may regret that Chinese astronomers were unable to provide us with a photoelectric magnitude of the supernova they observed in 1056, but what they observed and transmitted is the only datum available for studies of the variation of brightness of this supernova, which is now visible as the Crab nebula.

A further difference, which is obvious when comparing with data in other sciences, such as botany, is that astronomical objects outside the solar system are only accessible through their electromagnetic signals or the high energy particles they emit: we cannot touch, weigh or chemically analyse them, nor cut them or observe them under different illumination and angles. In short, the astronomer is a spectator of what happens.

3.2   Models and theories

Observations can only be converted into data by means of an interpretation, which may also be called a model or theory. Sometimes the relation is simple, like when we measure the radial velocity of a star through the study of the Doppler shift of the spectral lines. If we had no idea of the existence of a Doppler shift, we could only conclude that lines have different wavelengths in stellar spectra compared to laboratory spectra.

At other times the distance between observation and datum is rather remote, because the model is fairly complex. Two examples serve to illustrate this.

In 1919, during the phase of totality of the solar eclipse, British astronomers photographed the stars visible around the Sun and measured their positions. The models in this case were Newtonian mechanics and twentieth-century physics; they obtained a deviation between 'predicted' and 'observed' star positions, in agreement instead with the predictions of Einstein on the curvature of space. Here one sees that the distance between observation and datum is already rather large.

Another example is furnished by the determination of the datum 'average density of the universe within a sphere of a radius of one megaparsec'. To derive this datum one needs a knowledge of all galaxies within this sphere and their individual masses and positions in space. Each of these items is a datum in itself, derived from other data. But we would be at a loss trying to connect the datum 'average density' to any concrete observation. This is simply because the model that we use to derive this average density is very complex.

It is thus clear that our definition of datum as 'any parameter that can be deduced from observations by a convenient model' links very closely – almost inextricably – the models, the observations and the data.

The consequence of this relationship is that data only make sense in the framework of a certain theory or model. Furthermore, if the model does not propose a method to produce a datum from a given set of observations, no datum can be provided.

As a first example, reconsider the 1919 eclipse data: if Einstein had not predicted that a deviation should exist, observers would probably not have noticed the phenomenon, or if so, would have explained it away as a 'systematic error'. In such a case, one had the observations but, because of the lack of a model, would not have obtained the datum 'deviation of light near solar limb'.

To illustrate a second case, consider Lucretius (*c.* 70 BC). He spoke of atoms as the ultimate constituents of matter but he did not propose an experiment or observation from which he could derive their dimensions. This had to wait until the nineteeth century – then we got detailed models and could derive the datum 'average dimension of an atom'.

One may say that 'data are crystallized knowledge of a certain epoch' and this summarizes very well the situation, because it puts data in their historical context. Please notice that this is not true for observations, which may remain valid long after the time a datum was derived from them.

Let us take the Earth–Moon distance (d). When first determined by Aristarchos of Samos (*c.* 250 BC) the datum $d = 19K$ was obtained where K is the diameter of the Earth. Newton had the datum from better methods and used $d = 20K$ whereas, today, we accept $d = 30.13K$. Each datum is the fruit of a certain historical context, and should never be considered outside it.

3.3   Data in teaching

Usually, a student of astronomy is taught about theories and instruments – that is, how to make observations and how to interpret them. Little emphasis is placed upon data, because he is told that data are obtainable from the literature. When he becomes a professional it will turn out very often that he works with data and for data and that it is his knowledge of the quality of data that makes him a professional. Outsiders from other sciences – chemists, for instance – cannot judge the quality of the data and so we must make these available to them, in the same way as we take it for granted that they can provide us with more reliable values for chemical data than we could ever expect to get ourselves, even if we read the chemical literature.

3.4   Types of data

The values one gets from an observation are called *raw data*. According to what was said in Chapter 1, raw data includes information regarding the instruments with which the observation was made. So, for instance, if I wish to measure the position of a star on the sky, I must not only specify its altitude (or zenith distance) and its azimuth, but also the time, date, place, temperature and atmospheric pressure (for the refraction correction), when the observation was carried out. Each one of these items may consist of various readings – for example, the azimuth was read four times, the thermometer twice and so on. Obviously among these raw data there are some that are more important than others and some that bear a direct relation to the final datum I need (position on the sky), but we call them all 'raw data'. Its existence and preservation is essential if one wishes to re-use the data with, for instance, an improved model.

One can next correct anomalies and remove known instrumental biases and this leads to *calibrated* or edited data. In our example, if four readings of the azimuth were made and one differs very much from the other, one can remove it. Such removal is permissible if certain rules are obeyed, which we shall consider later in more detail. Temperature and pressure are used to calculate the refraction correction. We are left then with a smaller set of data, which summarizes but does not replace the raw data: these are the *calibrated* or edited data. The transition from raw to calibrated data is more complex when the observations are themselves complex. In an observation from space, for instance, instrumental biases may change with time and often this becomes visible only after some time; in this case the whole 'calibrated data' set must be calculated anew with the improved instrumental correction applied to the original raw data.

On the basis of the calibrated data of an observation we can pass next to the *reduced* or analysed data, expressed in physical or astronomical units. For instance, I can deduce the position of the star on the sky expressed in right ascension and declination at a given time. Or, in the case of a space experiment, I can convert the calibrated data, say, into flux per unit wavelength. Please observe that here again one must use a well-defined physical model. If the model proves afterwards to be incorrect, one must be able to go back to the calibrated data in order to redo the analysis.[1]

Is this now the end of the chain? That depends upon the interest of the astronomer: perhaps the star's position is a piece of information needed to derive the proper motion or the precessional constant. Or, in the case of the space experiment, the 'flux per unit wavelength' is one piece of information needed to construct a model of the stellar atmosphere.

Probably, the terminology of raw, calibrated and reduced data sounds unfamiliar and a bit pretentious. However, these distinctions are by no means trivial, even if one thinks only about the amount of information relating to each kind of data. In our example of a star position we had seven groups (each one containing perhaps four different readings of the scales). Then we had seven calibrated data and two (final) reduced data (right ascension and declination) at a given epoch. The difference is also reflected in the way data are presented.

3.5   Presentation of data

We have already pointed out the requirements for the presentation of observations, i.e. of raw data, in Chapter 1, so we may now pass to the presentation of calibrated and reduced data.[2]

First of all, the procedures used to carry out calibrations and reductions must be carefully described. As a rule, it is the simplest steps that are often unspecified, which nevertheless makes re-reductions very difficult. For instance, discordant data are often discarded, but it is very seldom that somebody specifies what is meant by 'discordant'.

The procedures can be described either in detail or, if the reduction technique is standard, by reference to a paper where the technique is described. Such papers must be fully quoted; a statement that 'the data were reduced according to standard procedures' is meaningless if one does not know what 'standard procedures' are. Those of 1920? Of 1930? The omission of a reference for the reduction technique is fatal for re-reduction: the published data cannot be improved any more, even if one knows that they are flawed. For reduced data, two more points are essential, namely the units and the errors.

3.6   Units, errors and standard values

3.6.1   *Units*

Units must be indicated clearly; without them a datum is incomprehensible since 48 may stand for 48 pc, 48 km/s, 48 mm etc. This seems obvious, but it is often not done. Look, for instance, at papers on photometry and try to find if the magnitudes quoted are derived from fluxes expressed in wavelength or in frequency units!

If possible, data should be given in SI (International System) units, except if astronomical use indicates otherwise (astronomical unit, solar mass, etc.). One reference book for the definition of units is the National Bureau of Standards Special Publication 330 (1977).[3] A more recent publication is *Quantities and units of measurement* by J. V. Drazil (1983). For astronomers, the handiest summary is probably *Astrophysical Quantities* by C. W. Allen (1973). An IAU publication, the *Astronomers handbook*, edited by J. C. Pecker (1966), also provides a list of constants, with emphasis on those used in astronomy.

3.6.2   *Errors*

Errors must be indicated because they show the quality of data and are of prime importance when the datum is used for comparison with theory. Errors are two types: *external* and *internal*. Differences among repeated measurements of the same datum under similar conditions arise from a multitude of small unperceived changes in the object, instrument and/or (in astronomical context) in the medium between the object and the astronomer (for instance in the Earth's atmosphere). Usually these differences are assumed to obey a Gaussian (also called normal) distribution. With this in mind, one can assign to each observation an (internal) error, which represents the dispersion of the observations around the mean value. To characterize the internal error one may use different quantities. Although given in any book on experiment design or elementary statistics, let us list them here, just for completeness. Assume one has measured x_i ($i = 1, 2, \ldots, n$) values, which constitute a sample of the (infinite) population of possible x values. We define the sample mean by

$$\bar{x} = n^{-1} \Sigma x_i,$$

which estimates the population mean, and the sample standard deviation by

$$s = [(n - 1)^{-1} \Sigma (x_i - \bar{x})^2]^{1/2}$$

which estimates the population variance. The standard deviation of the mean is given by

$$s \cdot n^{-1/2}.$$

Frequently one also comes across the use of the probable error of the mean, which is

$$0.67 \, s \cdot n^{-1/2}.$$

It is of course necessary to indicate unambiguously which one is used to characterize the internal error.

External errors, also called systematic errors, are much more elusive. It often happens that the same quantity measured under different conditions, with different methods and by various scientists, differs much more than internal errors would suggest. This is, as Eichhorn and Cole (1985) have shown, due to an incomplete modeling. Assume that the model used to convert observations to data includes five variables and that our datum is the flux as a function of wavelength. Consider now another procedure for determining this same parameter and assume that the model used also has five parameters. The measurements should provide identical answers – if the five parameters were sufficient to specify a realistic model. Assume now that another factor – for instance, the temperature of the receiver (T) – was neglected; for the sake of argument, assume that the first series of measurements was carried out at $T = -5°C$ and the second at $T = +10°C$. Mathematically speaking, we should have used a model with six parameters instead of five. Since we did not do that, the results of both series can differ by an unexpected amount, which we would call a systematic error and which is simply due to the (neglected) sixth variable.

Hunting down the discrepancies between series of observations reduced with different procedures is a complex but rewarding task, because it reveals the incompleteness of the model used. But this can only be done if a given parameter (in our case the flux as a function of wavelength) is determined by at least *two different* procedures. (If one uses the same procedure, one gets only an internal error.) Conversely, any result based upon one single method *may* be subject to 'systematic errors' (due to incomplete modeling) of which we are unaware. Because of this, one must be very cautious about error assessment.

In practice, both the external and internal errors are underestimated by observers and, as a rule of thumb, one can say that errors are at least twice as large as authors indicate (or admit).

Readers very often assume that a quantity given in a publication is accurate to the last figure quoted, so that a distance to a galaxy given as 585 kpc is accurate to within ±1 kpc. Unfortunately, when one looks at the error, one finds it to be of the order of at least 10%, which implies at least ±58 kpc. Quoting too many figures is a very unfortunate practice, which produces much unnecessary confusion.

The specification of errors becomes crucial when data from different sources are combined to get a *'best' value*. Let us start by observing that a combination of different data cannot be done exclusively at the 'reduced data level'. The analyst must know the reduction procedures used to convert from raw to calibrated and then to reduced data and, specifically, the conditions under which 'discordant' data were rejected. Some decades ago it was considered quite acceptable to reject measurements that gave 'impossible' results. For instance, if an observer of trigonometric stellar parallaxes came up with a negative parallax value he rejected it from the series of other measurements because a negative parallax is impossible by definition. Such a practice leads, however, to a systematic bias. It is clear that, if a small positive quantity is measured, which is affected by a relatively large error (e.g. $0''.005 \pm 0''.008$ r.m.s.) inevitably, in a long series of measurements some values are found that are negative. If one rejects them, one is pushing the average to a more positive but biased value and reducing the internal error. The rejection of 'discordant' observations is a subject around which a long literature has developed (e.g. Branham (1986) and references therein); part of the difficulty stems from the fact that one can reject discordant observations on either statistical grounds, or physical grounds. In the first case, one puts faith in the applicability of a stastistical theory, whereas, in the second case, one considers that there may be an instrumental or human error (or even an error in transcription) that permits rejection. Attention should be paid to the 'robust estimator' techniques developed by statisticians for cases in which the applicability of normal law may be put in doubt.[4]

When different data are combined to obtain 'best values' (see Chapter 9) the reader should get acquainted with modern mathematical methods.[5] It is rather unreasonable not to use well-tested techniques that are easily available. 'Best values' are called 'critical data' by physicists, but this name is rarely used in astronomy.

3.6.3   *Standard values*

If a given datum has well-determined internal and external errors, it is often chosen as a 'standard datum' – for instance the flux from Vega (α Lyrae) or the solar constant.[6] Standard data are the results of much work and are necessarily few. One specific difficulty arising in astronomy is that practically all data are variable with time and part of the effort to establish the standard datum goes into the study of possible variability. Both cases quoted – the one of the solar constant and the flux of Vega – have a long history of claims of suspected variability and the most one can say is that, with the level of accuracy reachable today, the variability is not evident. Standard values are important cornerstones for further measurements; they are used to check the consistency of reduction methods and the appropriate functioning of instruments.

Exceptions to the definition of standard as given above appear when 'standards' are selected by definition. This happens, for instance, with the spectral type of a few fundamental stars, such as the Sun, whose type is G2 V. In principle, there can be no objection against the adoption of one object as a standard, because it fixes essentially the 'zero' of the scale. But fixing various points by definition can involve risks because n points may generate disagreements. It is like defining the position of a curve with n (say 6) points: if the curve is a straight line, one is bound for trouble. Besides the two above-mentioned meanings, the word 'standard' is also used in a wider sense to denote a value that has been measured repeatedly (without establishing external errors). Such standards can be numerous and are called 'secondary standards' to distinguish them from the 'primary' or 'fundamental' standards defined already. The need for secondary standards arises, for instance, when one observes photoelectrically in any system – assume UBV, just to make it simple. A photometric system is defined through the choice of an instrument (aluminium coated reflector), a receiver (a photomultiplier sensitive to the 300–800 nm region), a set of filters (U = ultraviolet, B = blue and V = visual). One needs also one standard at least to define the zero point of the magnitude system. Obviously it would be nice to have a few stars around to check whether everything (telescope, receiver, filters, reductions) works properly and that is when secondary standards come in. Since one needs stars of different magnitudes to check the scale, stars of different spectral types and reddenings to check for colour effects and stars visible from different places on the Earth and at different seasons, one ends up with some hundred stars (for the UBV system).

Such a large number of stars can never be converted into fundamental standards, and this is also unnecessary: the stars are just 'cornerstones' to check the internal coherence of the measurements. It does *not* imply that these stars are 'well-behaved', except for UBV photometry; all that is required is that they be non-variables down to the precision of the UBV system (± 0.01 mag to ± 0.02 mag). Thus, a spectral type secondary standard is not necessarily a secondary photometric standard and, in fact, some MK standards are photometric variables. Standard values are collected in special works: we refer to Allen (1973) and Davis Philip and Egret (1985).

Matters of calibration and standards for stellar astronomy were considered at length in IAU Symposium No. 111 (Hayes, Pasinetti and Davis Philip 1985), where the reader may find very detailed discussions.

Notes on Chapter 3

For data in sciences see Rossmassler and Watson (1980). The terminology of raw, calibrated and reduced data was from Ochsenbein (1986).

1. One of the many books on data analysis is by Meyers (1975).
2. A guideline for the presentation of astronomical data is provided by Wilkins (1982).
3. Units are defined by international conventions (Conference Generale des Poids et Mesures) and are given in Bureau International du Poids et Mesures (1985). The choice of units is in the charge of the International Union of Pure and Applied Physics (IUPAP) (see Chapter 13).
4. Errors are discussed, for instance, in Rossmassler and Watson (1980).

5. For a survey of different statistical techniques see ESA (1983), which provides a large bibliography of mathematical textbooks. A recent book that covers part of the methodology is Murtagh and Heck (1987).
6. The variability of Vega (α Lyr) has been the subject of many discussions; see, for instance, Fernie (1981).

References

Allen, C. W. (1973) *Astrophysical quantities*, The Athlone Press

Branham, R. L. (1986) *QJRAS* **27**, 182

Bureau International du Poids et Mesures (1985) *Le système international d'unités (SI), 5th edition*, Bureau International du Poids et Mesures

Davis Philip, A. G. and Egret, D. (1985) Microfiche inserted in *IAU Symp. 111*. (See Hayes, Pasinetti and Davis Philip 1985.)

Drazil, J. V. (1983) *Quantities and units of measurements*, Mansell

Eichhorn, H. and Cole, C. S. (1985) *Celestial Mechanics* **37**, 263

ESA (1983) *Statistical methods in astronomy*, ESA SP-201

Fernie, J. D. (1981) *PASP* **93**, 333

Hayes, D. S., Pasinetti, L. E. and Davis Philip, A. G. (ed.) (1985) *IAU Symp. 111, Calibration of fundamental stellar quantities*, Reidel

Meyers, S. L. (1975) *Data analysis for scientists and engineers*, John Wiley and Sons

Murtagh, F. and Heck, A. (1987) *Multivariate data analysis*, Reidel

National Bureau of Standards (1977) *The international system of units (SI)*, National Bureau of Standards Special Publication 330, US Government Printing Office, Washington DC

Ochsenbein, F. (1986) In *Data analysis in astronomy, II*, Gesu V. di *et al.* (ed.), Plenum Press, p. 305

Pecker, J. C. (ed.) (1966) *Astronomer's handbook, Transactions of the IAU XIIC*, Academic Press

Rossmassler, S. A. and Watson, D. G. (ed.) (1980) *Data handling for science and technology*, North Holland

Wilkins, G. A. (1982) *Guide to the presentation of astronomical data*, *CODATA Bull.* **46**, CODATA, Paris

4

Archiving of observations

In Chapter 3 we have dealt with data in general. In this chapter we shall deal with the general problem of data preservation, or archiving. Since this is a rather vast subject, we shall divide matters in two parts. The present chapter deals mostly with the archive of observations, i.e. of raw data. We shall see that, in many cases, one does not preserve raw data but, rather, calibrated data – so the present chapter will include both. In contrast, the presentation and preservation of analysed data will be the subject of Chapter 5.

4.1 History

Let us start with some historical background on what archiving meant at different epochs of astronomy. The first astronomical observations were made visually and so the archive consists (for instance) of an inscription on a clay tablet: 'Venus seen the 6th Abu in the east in the constellation Ku Mal (Aries)'. If we still know where the observations were made, the record is complete. At a later time, objects were marked on a celestial globe and distances measured on it; still later, observations were made with better instruments, like gnomons and astrolabes. We do not know, except in a few cases, which observations were made with which instrument and the instruments themselves have not survived – two facts that prevent a re-reduction of the observations and critical evaluation of their accuracy. From the whole classical period of astronomy in Greece and Alexandria very few observations are preserved and what is preserved is contained mostly in Ptolemy's book called *Syntax* or (more usually) the *Almagest*. The total number of observations is certainly smaller than those we have from Mesopotamia; the latter are favoured over the former because archeologists discovered well-preserved libraries of clay tablets. Papyrus is more difficult to preserve and relatively few of them are preserved in readable form.

The fact that the number of observations is small has been explained in two ways. If few observations were carried out, because observations were not considered crucial as they are now, then their lack is natural and would correspond with the lack of data in other sciences like physics. If, however, the observations were made but lost or destroyed afterward, this would be in line, for instance, with the loss of libraries (e.g. those of Alexandria) and the loss of the clinical archives that we know existed at the medical school of the island of Cos. Probably both explanations must be combined but the fact is that we have no archive of observations left from classical astronomy in Greece, Rome and Alexandria.[1]

We are slightly better off in the case of Muslim astronomy, which was also nearer to our time, but the complete change came with Tycho Brahe (1546–1601), who made a determined effort to note his observations, to describe the instruments with which they were made and to store both. Writing up the observations made – i.e. keeping a log book – became an essential feature of all observatories in the next centuries. These log books were the first data archives in astronomy. We still have these documents for all the observations carried out with refractors and meridian circles during the following centuries and, thanks to these, it was possible, for instance, to re-reduce the observations of Bradley (Auwers, 1912–14).

With the use of photographic plates in the second half of the nineteenth century, a very powerful device became available to astronomers. Each surface unit of the plate reacts to the electromagnetic signal (photons) received and the effect integrates with exposure time. When adequately developed, fixed, washed, dried and stored (i.e. processed) the plate remains unchanged over time. Thus it preserves the electromagnetic signal received at the moment of exposure. In this way, photographic plates carry information and memory in the same place.

Experience has shown that plates can be preserved for up to a hundred years or so. In fact, one of the major projects of today's astrometry is to re-use the plates taken for the *Catalogue Astrographique* and *Carte du Ciel* 50–100 years ago.

To archive plates, observatories began to use plate vaults; we shall examine them later in more detail.

The next major development in instrumentation was photoelectric photometry. Here one got the output, which is proportional to the number of photons, in the form of a pen-and-ink record – i.e. a roll of paper on which the output was registered, as a continuous record. The archive in such a case consists of a collection of paper rolls.

In the second half of the twentieth century, techniques had new spectacular developments when charge coupled devices, both one- and two-dimensional, came into existence. Essentially, such a device consists of a number of isolated elements (pixels), which register the instantaneous electromagnetic signal falling on them. We see an instantaneous (or an integrated) image of the object and the result of each pixel is recorded on a number-counting device from where it can be transferred to magnetic tape. These devices are of higher sensitivity than photographic plates but, in contrast to these, the 'information' and the 'memory' are separated. The archive is constituted by a collection of the magnetic tapes on which the pixel records are stored.

Storage on tape is also necessary in two other cases. If the amount of information received is very large, for instance in the case of a radio telescope scanning the sky at a certain range of wavelengths, the records of observations are no longer made on paper rolls, as for photoelectric observations, but directly on magnetic tape.

If the information is gathered above the Earth's atmosphere by space vehicles we have in many cases the physical impossibility of bringing the observations down to Earth. The information is transmitted from the satellite by telemetry and stored directly on magnetic tapes or discs.

4.2   Support media

Raw data as well as calibrated and/or analysed data can be recorded on at least three types of support media, namely:

(a) paper, which has a lifetime of several centuries;
(b) photographic material, which lasts at least one century;
(c) magnetic tape or discs, which last a decade or so.

This does not include other kinds of support that were used earlier in history, like clay tablets, parchment and papyrus sheets; the reason is that such documents, to be understandable to us, have to be transcribed on paper.

Each of the lifetimes quoted above merits a comment. Although it is generally true that documents printed on good paper last several centuries, written documents may remain legible for much less time, depending on the quality of the paper and the way of writing. The shortest lifetime is that of photocopies (Xerox type); then follow manuscripts written with modern inks, machine-written documents and, finally, printed documents. As archivists know, a lifetime of several centuries is definitely not the fate of all documents on paper![2]

Similar observations apply to photography; everyone knows the fading of old pictures and of (recent) colour prints. A lifetime of one century applies only to emulsions on glass plates, carefully developed, fixed and washed, and adequately stored afterwards. Microfilms under the same conditions may have a similar lifetime but we do not know yet. Their practical use dates only from the 1930s. Finally, for magnetic tape, the lifetime depends also upon the way the tape was stored afterwards and lifetimes as short as three or four years may be more realistic than a decade.

The next device to be used for storage will probably be the optical numerical disc, whose use is starting now. Because of its novelty, its lifetime is as yet unknown, but may be longer (30 years?) than that of a magnetic tape. (For a survey see Gaillard, 1986.)

The only simple way to prolong the lifetime of data presented on a certain type of material is to copy them anew, before degradation sets in. Old books, like Ptolemy's *Almagest*, were copied and afterwards printed several times over the centuries, whereas the originals have perished long ago. Similarly, magnetic tapes can be recopied every few years to prevent degradation and preserve their content. The problem is more complicated for photographic material. Preliminary experiments on one-century-old plates have shown (Rousseau and Guibert, 1985; Colin and Soulié, 1988) that degradation is not yet serious but we do not know if this is the general rule. Presumably, a copy of the (aged) original may not be a satisfactory substitute.[3]

4.3   Data archives

We define a (data) archive as an ordered collection of documented data, collected at a particular observatory and organized in a way to permit permanent storage and retrieval. Let me comment on the different elements in this definition.

(a) Data. Notice that no distinction is made between the three kinds of data (raw, calibrated, reduced); 'raw data archives' are also called 'observational archives'. Usually, raw data or calibrated data archives are mainly sought by specialists in a given area, whereas non-specialists in the field look for reduced-data archives.

(b) Ordered collection. The first condition a data archive must fulfil to be usable is that it be ordered. To facilitate retrieval the order should be such

as to permit various selection criteria but this is really secondary compared with the requirement that it be ordered.

(c) Data documentation. We have already insisted in the previous chapter on the need to document all data carefully. In an archive that is attempting to preserve data for further use, the lack of documentation is equivalent to the loss of the archive.

(d) Collected at a particular observatory. This applies fully to observational archives; for instance, we have the Observatory of Haute Provence plate archive and we find there only material obtained at that observatory (and not at some other observatory). For reduced data there exists more latitude and sometimes archives contain data collected at other observatories, as happens, for example with the 'data depositories'. This characteristic of archives, that they are related to observations made at a given institution, is what distinguishes data archives from museums and libraries.

(e) Permanent storage. We have already seen that 'permanent' means in this context 'over the duration of the physical support' – several centuries for paper records, several years for magnetic tape. Special precautions have to be taken to guarantee the best survival of the archives, such as protection against theft, vandalism, fire, water, atmospheric pollution, humidity, dust and/or strong illumination. The reader can find some references to papers dealing with these technical aspects in the notes at the end of this chapter.

(f) Retrieval facility. To be useful, an archive must be organized in such a way that retrieval is not only possible, but also easy. Since archives are usually bulky, this is usually achieved in two steps. First, one does not look for the datum itself, but only for its existence. If the datum exists, one may then retrieve it.

The first step consists of a list of observations, which is part of the ancient 'observing book' or 'log book'; we shall call it 'observing file'. The organization and content of this file depends essentially upon what is archived and so varies from case to case. Let us take as an example a plate archive. One may envisage retrieval according to plate number, to the position of the object (or of the region photographed) and/or the observing epoch. Additional information which could (or should) also be available in the observing file concerns the quality of the plate, exposure time (and/or the limiting magnitude reached), the instruments used (telescope, camera, spectrograph) and, if the plate has been loaned, to whom.

It is very clear that archiving is a rather complex task, which requires a relatively large effort. Such an effort is only justified if one has a clear idea about the central problem, namely: (a) why do we want to preserve data and (b) what type of data should be kept?

4.4   Why do we preserve data?

The basic reason is that practically all astronomical data are variable with time; observations made at a past moment are thus irreplaceable and have to be stored in memory. There exists an often-expressed criticism of this, which consists of saying that, 'usually', phenomena have sufficiently short periods, that more precise modern observations may be substituted advantageously for the older, presumably less precise, observations. The difficulty lies in the meaning of the term 'usually' in this connection. The lightcurve of objects such as η Carinae, the possible recently-formed craters on the Moon, the length of the day, the Earth's precession, all fall outside the 'usual' range. Besides this, we do not know for what purposes the next generation of observers will use our data. It seems inescapable that we have to keep the old data in the 'memory' of the astronomical community.

4.5   What should be preserved?

The second question is much more difficult to answer. As I have remarked already in Chapter 3, the amount of information needed to store raw data is much larger than that needed to store reduced data, and the amount of storage needed for analysed data is still smaller. So, one could be tempted to say that one should only keep analysed data. Before going into details let us quote some figures on the amount of data generated by some instruments. Table 4.1, taken from Grosbol (1982), provides some figures.

To obtain the total data flow, we must multiply such figures by the number of telescopes working (let us assume 100) and by the number of

Table 4.1. *Data acquisition rates in astronomy*

Sources	Mbytes/frame	frame/night	Mbytes/night
Schmidt plates	1000	5	5000
9-cm McLullan camera	100	20	2000
Prime focus plates	500	10	5000
CCD camera	1	50	50

nights per year (let us assume 200). If photographic plates are included, this results in about 10^{14} bytes = 100 Tb per year (1 Terabyte = 10^3 Gigabytes = 10^6 Megabytes = 10^9 kilobytes = 10^{12} bytes). The satellites add some more data. IUE, for instance, sends down 10^{11} bytes/year and the Space Telescope is expected to send 1 Tb/year.

Clearly these figures represent only an order-of-magnitude estimate. On the other hand, Ochsenbein (1986) estimated an output of 6 Tb per year for all ESO (European Southern Observatory) telescopes. If we assume that there are 10–20 observatories of the size of ESO in the world, we arrive at a total of 60–120 Tb/year, which agrees with our estimate.[4]

One may lower such estimates by a large factor, taking into account that photographic plates do not necessarily need to be stored on tape, because the memory is located on the plate. The figure of 100 Tb can then be reduced to a few Tb per year. With present magnetic discs, with a storage capacity of some 5×10^2 Mb, however, even such an amount represents a large number of discs.

On the other hand, it has been found that modern photographic emulsions degrade more easily and rapidly than old material and, in such cases, it would be better to scan the plates immediately 'to preserve the observations' – provided the (optical) disc has a longer lifetime than the (undegraded) plate. Besides the storage difficulty, it is clear that, if observations are put on tape, they are more readily accessible than if they are kept on the plate.

If one has difficulties in storing everything, one must examine carefully what kind of data (raw, reduced or analysed) should be preserved.

Let us consider a rather simple case as an example, namely photoelectric photometric observations. As already remarked, thirty years ago, the output of the amplifier was recorded by a pen on paper. Intensities were then read off and sky corrections applied. Standard stars observed for extinction were analysed next to derive night extinction coefficients, and the resulting extinction corrections were then applied to all observations, so that one got the magnitudes and colours for each star for this night. Archiving raw data means keeping the rolls of pen-and-ink records in drawers, together with the hand-written reduction sheets.

The process is now being done on many telescopes with microcomputers in such a way that the original photon counts can be converted, the same night, into magnitudes and colours. This means that, for such measurements, the raw data need not be kept any more and are replaced by analysed data.

A little reflection shows that this is due mainly to the fact that the procedure for converting from raw to analysed data is sufficiently well know in all details, that a new reduction of the raw data is not foreseeable. Similar considerations can be applied to astrometric data observed with automated meridian circles. However, even here some observers prefer to keep all raw data for a certain period, say ten years, so as to make a re-reduction possible in case of necessity. In both cases, full details of the reduction procedure must be published, so that future generations may know what we did in the 1980s, and how we did it.

An example of the opposite situation is the case of observations made from satellites. Since each satellite experiment represents an innovation, it is clear that the reduction procedures to convert from raw to calibrated, and later to analysed data, are subject to improvements. To take the case of the IUE satellite, the first calibration procedures left out one parameter, which later proved to be important (the thermal correction of the calibration file). The reduction procedures had to be modified, implying that all the data already outlined had to be reduced anew. One sees that, if the raw data had not been preserved, this would have been impossible and the calibrated data obtained before the change in reduction procedure would have been lost.[5]

The conclusion seems obvious: if the procedures to convert from raw to analysed data are well-mastered, the need to keep the raw data diminishes sharply and one can keep the latter only. It is, of course, difficult to lay down a rule specifying after how many years the procedure is 'well-mastered', but a few decades seems a reasonable time. Because analysed data occupy a much smaller volume than raw data, analysed data can then be printed and stored on paper, which preserves them at least over several centuries.[6]

4.6   Types of data archives

We shall analyse next four different types of data archives, which illustrate different aspects of the archiving problem. We shall consider in turn plate archives, the archives of a (national) observatory, the archives of space experiments and the case of a reduced data archive.

4.6.1   *Plate archives*

Both the technical and the administrative aspects of plate archives have been discussed a number of times.[7] Oversimplifying the problems, one might say that, whereas the technical aspects, like the optimum room temperature and air humidity of the plate vault and the material of the envelopes, are well known, the organizational aspects are still in a state of flux. Plate archives were established first for the needs of the staff members, to keep track of what had been observed. Thus, the preservation of the plate, plus an entry in the observing book (log book) of the object or the region photographed was adequate. When plate collections grew to several thousands, usually a (hand written) card file was created to supplement the log books.

Problems arise very soon with visiting astronomers from other institutes: if they take plates, should they be allowed to carry them away for further analysis? Such questions became, of course, pressing at the national observatories. Over the years, a kind of consensus evolved that plates taken by visitors should be considered the property of the observatory where they were taken, but loaned to the astronomer who took them. The 'loan time' varies from place to place, but is usually of the order of a few years. (Administrators of plate archives usually complain that nobody returns plates before his retirement.) Exceptions to such a policy arise when the plates taken are difficult to carry away – like thin Schmidt plates. In such a case, observatories make available only contact copies of the originals.

Difficulties arise when one wants to know whether a certain object or a certain region has been observed, i.e. is on the observing file. For example, one might like to know if the spectrum of a nova before its outburst exists in the archives. Since card files were usually organized before the computer era, they are often ordered in chronological order, or only in a position-on-the sky order (by increasing right ascension for instance), but seldom by both. Rather rarely, observing files have been put in machine-readable form, as an enquiry has shown (Hauck 1981).

Hauck found that, out of 28 plate archives consulted, with a total of 1.2×10^6 plates involved, only eight had an observing file usable by computer (in 1980). The explanation for this is that such an operation is often very expensive, if the plate archive is old and voluminous. Only at recently-installed photographic telescopes has this been done from the start (see for example Tritton (1982) and Heudier (1985)).

The difficulties are even greater when one needs to consult more than one archive. In such a case one has to write to each institution separately. This is a situation where data are stored but are inaccessible, because nobody knows what was observed and where.

A solution to such problems is not easy, because it requires cooperation between different institutes, which may not even have computerized observing files. One can only hope that cooperation will be organized in the future, probably through the existing data centres, which could also act as clearing houses for the observing files (Hauck, 1981).

4.6.2   *Observatory archives*

The difficulties encountered with a plate archive are magnified when we move to the archival problems of, say, a national observatory. In such establishments, a number of telescopes are used with a large variety of sophisticated auxiliary equipment – spectrometers, spectrographs, photometers, direct-imaging devices (e.g. plates and CCDs), polarimeters and so on. Most of this equipment is not attached permanently to any telescope but is changed, often after a few days. The output of the detectors is usually registered on magnetic tape, although listings, images and photographic plates are other possibilities. Astronomers using the instruments come for short periods and want to take back the results; in addition, they have no time to become familiar with the instruments during their (short) stay: detectors are more or less considered as black boxes by them.

It is clear that a number of difficulties appear when one considers the archival aspects. If observers take the material home, the archive becomes scattered over a number of places, with only a slight chance of being re-united again. In line with the policy for plate archives, observatories consider that the data collected are only loaned to the observer and wish the material to be returned after some time. In the case of data on magnetic tape, the problem may be partially solved by providing the astronomer with a copy of the original tape, so that the other one is available for the archive. Provision must be made for the purely technical data relating to the telescope or the proper working of the receiver also to be stored somewhere because, otherwise, any re-reduction of the observation may be jeopardized. As remarked before, the astronomer is usually not in a position to know exactly what happens inside the 'black box', so these data are of less value to him than to the technicians in charge of the operation of the system. Therefore, the astronomer is usually not provided with these data but only with those essential for the reduction of his observations.

The sum of all considerations has led in recent years to an overhaul of the old ideas of archiving, which is reflected in the set-up for the British facilities on the Canary Islands and for the European Southern Observ-

atory. Raimond and Lupton (1986) summarize the purpose of the system (for the Canaries) in such terms. 'The purpose of the system is to:
  (a) collect and organize raw data at the telescope;
  (b) transport the data to the machine that loads the archive;
  (c) load data into the archive, creating an observations catalogue and an archive index;
  (d) allow easy interrogation of the contents of the observations catalogue and the archive;
  (e) allow easy submission of retrieval requests, which will result in extraction of data either on disc or as FITS tape.'

To this very clear description of the objectives it should be added that the system is targetted at both astronomers and engineers. This, as we have seen, constitutes considerable progress because, in the old observatories, engineering data were kept separatedly (if kept at all) and are usually difficult to locate and use. The observing file will be permanently accessible on line, whereas archives stored on magnetic tapes are kept off line. It is to be hoped that the introduction of such a new approach will have a lasting impact on archiving practices, although smaller observatories will probably be less motivated or have less funds to spend on such a sophisticated system of archiving.

4.6.3 *Space experiments*
Parallel to such developments 'on the ground', space agencies started also to realize that archiving is an important part of their activities. Between the first satellites, the instrumentation changed and improved rapidly so that each new satellite was regarded as an instrument superseding earlier experiments. Furthermore, the number of data obtained by each satellite was not very large. Both factors characterize what might be called the 'experimental phase' of space observations, a phase in which archiving was not perceived as something crucial.

The situation has changed now and archiving must be regarded as an extremely important activity, to be organized in all details. Heck (1986) lists a number of requirements from which we quote a few:

  – an adequate survey, including simulations, should be performed *before* launch to select the appropriate information to be kept;
  – before the end of a mission, plans should have been made for a complete reprocessing of all the data collected with the latest and supposedly best version of the processing software;

  – (after) the end of a spacecraft life . . . (its) data base should be maintained with an appropriate service to the astronomical community.

Obviously the requirements must be adapted to each satellite and it is clear that the more recent experiments have benefitted from the experience of earlier ones. Let us also remember that, in a broad sense, satellites may be either of the survey type or of individual object type. As an example of the first kind we have IRAS (Infrared Astronomical Satellite), which scanned the sky in the infrared and produced, at the end of the mission, a huge catalogue of infrared sources. As an example of the other type of observations we have IUE (International Ultraviolet Explorer) in which isolated, preselected, targets are observed. Heck's recommendations (as he says himself) are particularly applicable to the second type of experiment, although the first and the last requirements listed apply to all types of satellite experiments.

Space Agencies have not yet considered the full problem of archiving because they think in terms of conservation over a decade or so. Although this is progress over times when they thought in terms of a few years, it is not enough for astronomers, who think in terms of a century or more. One partial solution would be to print the reduced data. Although this is considered to be expensive, it will cost only a tiny fraction of the satellite itself and will ensure a lasting archive of at least the reduced data. No easy solution exists for calibrated data.

4.6.4   *Reduced data archive*

  As an example of an archive not related directly to observations carried out at a given observatory, let us consider the case of the American Association of Variable Star Observers (AAVSO), which archives (since 1911) the visual observations of variable stars made by amateurs. The history of this archive is well summarized by Waagen (1986); the data influx is of the order of 1.5 to 2.0 $\times$ 10^4 visual observations per month coming from about 500 observers all over the world. The total number of observations archived is of the order of 5.5 $\times$ 10^6, which testifies to the large number of amateurs active in this field over the years. Since most amateurs are not used to publishing their observations and few journals exist that publish amateur observations, most of the observations made by amateurs would be lost if it were not for the AAVSO.

Amateurs send in their (visual) estimates on a monthly basis, using a standardized report form and reporting procedure. At the AAVSO, observations are checked and entered in the computer. The unprocessed 'raw' observations are also kept, to ensure future accessibility.

The preceding four cases illustrate some of the complexities of archiving and stress the fact that archiving is an activity that must be carefully planned from the start. Since observations are accumulating at a tremendous speed, any fault in initial planning rapidly affects a large number of observations; when the fault is corrected, usually the new solution is only applied to the most recent material and the previous material often must be left as it is. This is mostly due to the impossibility of providing some of the missing data or because funds and/or manpower are lacking for the re-ordering of the archive.

The main requirements for archiving perhaps may be summarized in the following way:

(a) define a clear policy of archiving, and adhere strictly to it, enforcing users to follow its procedures;
(b) impose strict standards for preservation of the information support medium (precautions against vandalism, theft, fire etc.);
(c) ensure that observing files be easily accessible, well organized and trustworthy;
(d) make provision for long-term archiving, even if at present this is not regarded as essential;
(e) also preserve the technical data on telescopes and receivers in such a way that re-reduction is possible.

Readers will probably be surprised by the insistence on organization because, after all, archiving is part of observing and should thus be well-mastered. Nevertheless it is true that many institutions have no clear archival policy and that archiving is very rarely discussed in the literature. It was the lack of such references that led me to organize a meeting on 'Archiving Astronomical Observations' at Montpellier. The transactions (IBCDS No. 31, 1986) provide the only recent systematic coverage of the subject.[8,9]

Notes on Chapter 4

1. The difficulties that appear when historians analyse ancient Greek observations is illustrated in the case of the Hipparchos catalogue of stars, which is only fragmentarily preserved. For an analysis of its content see Vogt (1925), Neugebauer (1975) and Nadal and Brunet (1984); the latter authors carry out an analysis of the (lost) instrument used for the observations.

2. For an account of difficulties with paper records, see Corbin (1988) in the astronomical context and Favier (1985) for archives in general.
3. For another more optimistic life-time estimate of the different kinds of support media see Calmes (1986).
4. Although the number of data being archived grows rapidly (see Chapter 8 for details), we are still a long way away from a situation like that of many historical or administrative archives where the volume is such that either material is selected according to its importance, and the rest destroyed, or sampling is done in the sense that, say, only the records of each fifth are preserved. Either decision will certainly be criticized by the next generation of archivists and/or historians!
5. The 'missing parameter' of IUE, which forced a re-reduction of previously-made observations, is given in NASA (1982).
6. That even in the best-mastered data-reduction procedures innovations may occur is indicated by the discovery of internal refraction (i.e. in the telescope tube) in meridian circles by Høg (1986). Meridian circle observations have been made for almost three centuries and it was thought that all the details of their reduction were well known!
7. I am indebted to Dr A. R. Klemola for the following references to plate archives:
   Klemola A. R., Vasilivski S., Shane C. D. and Wirtanen C. A. (1971) *Publ. Lick Obs.* **22**, part 2 (see p. 6)
   Van Altena W. F. (1972) *AAS Photo Bulletin* No. 2, p. 15 and No. 8, p. 18
   Burnham J. M. (1973) *AAS Photo Bulletin* No. 1, p. 12.
8. As an example of the little interest shown by the astronomical community in archiving problems let us say that neither the *Handbuch der Astrophysik* (Springer Verlag) nor the *Stars and stellar systems* series (Chicago University Press) contains specific sections devoted to archiving. It is not only necessary to have an archiving policy defined, but its application must also be enforced. It is easy to illustrate the necessity for this, if one considers simply the number of sloppy entries in observing files at any observatory (coordinates in unspecified equinoxes, object designations using unconventional names, omission of instrumental data, etc.), the straight lack of data (no trace of photoelectric observations carried out twenty years ago), the absence of engineering data (no trace for instance of the dates when a photographic triplet was dismounted, cleaned and re-assembled), or inappropriate storage (log books stored in humid cellars, plates stored in rooms under roofs where temperatures can be both extremely hot or cold, and so on). And then, of course, there are happy-go-lucky astronomers, like the one who scribbled entries intended for the log on the wall of the dome (so that a painter later obliterated all records), who do not care at all about archiving.
9. An enquiry on archiving will be published in *BICDS* **34**. It will give the answers from more than a hundred observatories that were consulted.

References

Auwers, A. (1912–14) *Bearbeitung der Bradleyschen Beobachtungen* Vol. 1 (1912), vol. 2 (1913), vol. 3 (1914)
Calmes, A. (1986) *Library Journal*, Sept. 15 issue

Colin, J. and Soulié, G. (1988) in *IAU Symp. 133, Mapping the sky: past heritage and future directions*, Debarbat, S. *et al.* (ed.), Reidel

Corbin, B. (1988) In *IAU Symp. 133, Mapping the sky: past heritage and future directions*, Debarbat, S. *et al.* (ed.) Reidel

Favier, J. (1985) *Les archives, 4th edition*, Presses Universitaires de France

Gaillard, C. (1986) *BICDS* **31**, 49

Grosbol, P. J. (1982) In *IAU Coll. 64, Automated data retrieval in astronomy*, Jaschek, C. and Heintz, W. (ed.) Reidel, p. 79

Hauck, B. (1981) *BICDS* **20**, 68

Heck, A. (1986) *BICDS* **31**, 31

Heudier, J. L. (1985) *BICDS* **29**, 19

Høg, E. (1986) private communication

Nadal, R. and Brunet, J. P. (1984) *Archive for the History of Exact Sciences* **29**, 201

NASA (1982) *IUE Newsletters* **16**, 67

Neugebauer, O. (1975) *A history of ancient mathematical astronomy*, Springer Verlag

Ochsenbein, F. (1986) *BICDS* **31**, 25

Raimond, E. and Lupton, W. (1986) *BICDS* **31**, 41

Rousseau, M. and Guibert, J. (1985) *BICDS* **27**, 43

Tritton, S. B. (1982) In *IAU Coll. 64, Automated data retrieval in astronomy*, Jaschek C., and Heintz W. (ed.), Reidel, p. 229

Vogt, H. (1925) *Astron. Nachrichten* **224**, 17

Waagen, E. O. (1986) In *The study of variable stars using small telescopes*, Percy J. R. (ed.), Cambridge University Press

5

Presentation of astronomical data

I shall deal in this chapter with the ways astronomers present data. Of the three categories of data, I shall consider in this chapter mostly analysed data; matters relating to the other kind of data were discussed in Chapter 4.

I have preferred to discuss matters of presentation here, although, alternatively, I could have first discussed the subject matter of Chapter 9 (data banks and data bases). If I prefer to discuss presentation now it is because I am following the historical order, since astronomers invented catalogues first, long before the idea of data banks was introduced. Because of this, Chapters 5 and 9 are complementary in several ways.

When one speaks of presentation of data, one implies also the existence of a support medium. Since most of the data appear in print, this implies that we use a paper support, which preserves data for a certain time, but not indefinitely (see Chapter 4).

5.1   Data in research papers

Let us examine now how analysed astronomical data are presented. A perusal of the astronomical literature (see Chapter 11) shows that most data are presented in research papers in scientific periodicals. On the other hand, the amount of data contained in a paper varies considerably, from one item to several hundred.

It is usual to call 'lists' data collections for more than, say, ten and less than one hundred objects and 'catalogues' if they are larger data collections. In general one finds that there are many papers providing a small amount of data, whereas there are far fewer lists and even fewer catalogues.

Ochsenbein (1982) has derived some statistics that illustrate this point. He studied the average number of stars on which data have been published in papers. His result was that, in the previous years (1975–78) this number

Table 5.1. *Frequency of papers quoting N data (for 1978)*

N stars	%	N stars	%
≥100	3.5	7	2.6
50–99	4.3	6	4.2
30–49	3.3	5	4.9
20–29	4.5	4	6.5
15–19	3.7	3	9.3
10–14	5.6	2	12.4
9	1.6	1	31.6
8	2.0	Total	100%

The first column gives the number of stars, the second the percentage of papers (out of 1476 papers).

had fallen, to about 20. More detailed statistics of the numbers of papers quoting data on stars is given in Table 5.1.

From this table one can see that 31% of all papers just provide data for one star, that 76% of all papers contain data for less than ten stars, 21% between ten and a hundred and only 3.5% contain more than one hundred stars. This illustrates perfectly the rule, 'many papers with few data, few papers with many data'.

A similar statistic was derived by Mermilliod (1988) on papers dealing with UBV photometry. In the 1419 papers dealing with this kind of photometry in the period 1952–1985, 87 267 stars were measured – the average is thus 60 stars. Detailed statistics show, however, that 31% of all papers contain less than ten measurements and 52% less than three. Because UBV is a very popular system, with well-known reduction techniques, one would expect the average number of stars to be much larger than in Ochsenbein's statistics, in which *all* observational papers are considered together. Since 60 is much larger than 20 this expectation is verified, but the range is still very large. There are 26 papers (i.e. 2%) that contain more than 500 stars, with only one providing 3000 stars.

Such a state of affairs obviously constitutes a drawback for anyone interested in a particular set of data, because it implies long and difficult literature searches. It is obvious that astronomers usually look first for catalogues because there are only a small number of catalogues but each one contains a large number of data. Results published in research papers tend to become bibliographically inaccessible after some time because of the sheer impossibility of examining a large journal collection just to find one datum. Such a situation has obviously led to various attempts to solve it, which I shall summarize. I will start by examining the concept of a catalogue.

5.2   Catalogues

A catalogue is a long list of ordered data of a specific kind, collected for a particular purpose. We shall come back later to this definition; for the moment I will quote some examples to make the terminology clear.

Argelander (1859–1862) in his *Bonner Durchmusterung* published the positions of 324 198 northern stars (declination north of −2°), derived from observations made by himself and his collaborators. The catalogue is properly called an observational catalogue (OC), because it reports data (positions) obtained by himself.

Jaschek *et al.* (1964) published a *Catalogue of stars classified in the MK system*. The majority of the classifications reported were not made by them – so this is a compilation catalogue (CC).

Jenkins (1963) published a 'catalogue of stars with trigonometric parallaxes'. Some of the observations (but not all) were made by her – so we have again a compilation catalogue. However, she also derives from the data compiled an average parallax, so she has added something to the data collected. This I designate as a 'critical compilation catalogue' (CCC), whereas the one by Jaschek *et al.* I call a 'bibliographic compilation catalogue' (BCC).

Finally, we have the *Bright Star Catalogue (Fourth edition)* by Hoffleit and Jaschek (1982). Here we have a general compilation of many parameters (positions, proper motions, radial velocities, spectral types, luminosity class, magnitudes, colours, rotational velocities, variability and binary system indicators) drawn from many sources; I call this a 'general compilation catalogue' (GCC).

Please observe that my terminology of OC, BCC, CCC and GCC refers only to the general type of information it contains, and not to specific characteristics. So, I do not include as separate types the astrometric catalogues providing 'absolute' or 'relative' positions, nor do I make a distinction between 'fundamental' and 'semifundamental' catalogues in the astrometric sense. These finer subdivisions characterize astrometric catalogues but are not very useful outside astrometry. In our terminology, 'absolute' and 'relative' catalogues are OCs and fundamental and semifundamental catalogues are CCCs.

The same remark also applies to the quality of the catalogues. For instance, among the bibliographic compilation catalogues a fine analysis will disclose a great variety of qualities: there are colleagues who did a careful job as there are others who did a sloppy job. The same is also true for the other types of catalogues.

Obviously, the four different types of catalogues have different purposes and address themselves to different users. Compilation catalogues address professional astronomers who want to have access to the original data. If one is an 'insider' in the field, one prefers BCCs because one knows how to use the data but if one is an 'outsider' one prefers CCCs because they provide one with a reputable scientist's opinion of what the best average value is. GCCs are preferred by all those who need a rapid idea on the data available. Notice, however, that for each parameter one needs to know in more detail one should go to a BCC or a CCC.

We may now define more precisely a catalogue as an ordered collection of a large number of data of a particular kind, collected with a particular purpose and obtained at a particular place and epoch. Let me next comment briefly on the different elements contained in the definition.

(a) Purpose of the catalogue. Because of its importance for the user, the purpose must always be spelled out clearly: it defines the aim that the author had in mind when observing or collecting the data. It also tells which are the data for which the catalogue was composed and these I shall call the 'primary data'. So, for instance, in a photometric catalogue the magnitudes and colour indices are the primary concern, whereas position and proper motions (if provided) are of secondary importance. In an astrometric catalogue exactly the opposite is true. The user should pay attention only to primary data, since secondary data are merely there for information or identification. The distinction is important because many older catalogues carry much secondary data – at a time when no computer-readable catalogues existed (because computers did not exist) it was easier to include all convenient information that could be crammed onto the printed page. The *General catalogue of stellar radial velocities* by Wilson (1953) has the main purpose of providing average radial velocities for stars but it also carries (as secondary data) positions (1950), spectral types, magnitudes and total proper motions. On the other hand, the catalogue

carries no bibliographic information on where the radial velocities come from. With hindsight it would have been more useful to include these and omit the secondary data. Another equally important objection to the inclusion of secondary data is that they acquire the same status of respectability as the primary data, even if they were selected in a rather uncritical (and generally non-documented) way – like Wilson's total proper motions.

(b) *Quantity of data*. I have already defined 'large' as meaning more than one hundred. Smaller data collections are called lists and not catalogues.

(c) *Order of the data*. For ease of retrieval, it is clearly better to order the data provided in some way in the catalogue. The commonest order is by equatorial coordinates (right ascension and declination at a given equinox) because this way one knows when and where the objects may be observed. Order by right ascension is used in most catalogues, although, from an observational point of view, order by right ascension in declination zones is preferable. The latter was used by Argelander for the *Bonner Durchmusterung* and the system was followed essentially in most catalogues of the nineteenth century.

Other possible orders are by constellation plus order of discovery (like for variable stars, where V 363 Sco means that it was discovered later than V 358 Sco), galactic coordinates, temporal order (Supernova 1056) and so on. (See Chapter 6 for more details.)

Let us notice that the 'ordering' usually refers to two different things – an 'internal' order in the catalogue (for instance a running number) and an 'external' order, which makes it possible to locate the catalogued object in the sky. We call these respectively the 'internal' and 'external' identifiers of an object. For instance, the internal numbers HD 85558 and HD 85563 identify two stars that are close together in the HD catalogue but, on the sky, they are far (38°) apart. Object identification and designation is so important that I devote the whole of Chapter 6 to it; I refer the reader to that part of the book.

(d) *Data explanation*. Obviously, data listed in catalogues should be carefully described. I refer the reader to section 3.5 (Presentation of data) for the basic requirements. I add simply that in the case of compilation catalogues, full bibliographic references must be given to the sources that were included in the compilation. If one cannot go back to the original sources, the catalogue is worthless and furthermore a flagrant plagiarism –

the original author who did the work is deprived of authorship in favour of the compiler.

If the catalogue is of the critical compilation type, the procedures used should be clearly explained or references be provided to papers where they are explained in detail. If possible, it is preferable to provide both the original data and the averages derived from them, so that the reader may see what was done.

Frequently, catalogues also carry 'notes' beside the tabulated information. Here again we repeat the requirement that 'notes' should be clear and non-ambiguous. If, for instance, a spectrum is peculiar, do not write 'spectrum peculiar' but say '$\lambda 4077$ and $\lambda 4215$ somewhat strong for luminosity class', which is at least clear.

(e) Location in space and time. Curiously this point is often neglected in OC (!) – we do not learn when the observations were obtained, a fact that diminishes their value for studies of time variability. Pointing out where observations were made may be crucial if one wants further information. (See also Chapter 1.)

The documentation of compilation catalogues also needs specification of the time interval covered by the data compiled. And in the GCCs, of course, it makes a difference if the parameters are taken from compilations done at different epochs.

Once a catalogue is printed it has acquired its definitive form, but still two things may happen, which we call the 'afterlife'. The first is the detection of errors in it and the second the preparation of a computer-readable version.

5.3   Errors in catalogues

Although every author wants to produce an error-free catalogue, it is a fact of life that, sooner or later, errors will be discovered – typographical errors, clerical errors, wrong data, wrong identifications (e.g. the data for HR 5999 confused with those for HR 6000). Even when an error has been detected it usually has a happy long life. No scientific journal is eager to print corrections, for one thing, and, on the other side, even if published they are largely ignored by users. In the last years the *Bulletin d'Information du Centre de Données Stellaires* has been publishing many lists of errata and it is hoped that this will be continued. It is important to notice that the corrections to errors *must* be published somewhere so that they can be distinguished from transcription errors, especially in the case of computer-readable catalogue versions.

5.4   Computer-readable catalogues

Such versions have become usual only in the last decade and the conversion from print to tape is not yet finished, since many catalogues have as yet no computer-readable version. Computer readable versions (CRV) of catalogues are popular because:

(a) data becomes easily accessible;

(b) data can be ordered and sampled in ways different from that in the printed version;

(c) catalogues can be easily updated and errors corrected;

(d) the computer-readable versions of different catalogues may be easily combined, retaining only the primary data of each catalogue;

(e) statistical studies (and more generally, any computation on the data) as well as graphical representation ('visualization') may be done easily.

There are, however, two disadvantages, namely:

(a) the short life span of modern magnetic tapes, which are one order of magnitude less durable than the printed versions. (See, however, Chapter 4 for possible improvements.)

(b) the catalogue becomes outside the control of its author: whereas printed versions are essentially immutable, CRVs are readily changeable by anyone who has access to the tape.

Ideally, the printed version (PV) and CRV of the catalogue should contain the same information but very often they do not. 'Notes' of the PV may have been left out, 'introductions' have been shortened and so on. What is the best policy? Should a CRV be a complete substitute for the PV or should it only facilitate easy handling of data? Many of the old catalogues being now out of print, one could probably favour the first answer. But what if there are hundreds of pages of notes? Since it is not obvious that the notes in a CRV are very easy to consult, it is probably best to adopt a compromise between both solutions – i.e. reproduce the whole catalogue but only a reasonable selection of the notes.

Another nuisance may come up in the sense that the person who prepared the CRV of the catalogue decided to alter details of the PV. If one only corrects errors (which have been published somewhere in print) this is fine, but what happens if one decides to suppress secondary information contained in the PV?

Although the problem seems trivial, there are four computer-readable versions of the catalogue AGK3 (Heckmann and Dieckvoss 1975), all derived from the same data set, as discussed by Fresneau (1981) and such a multiplicity of versions certainly does not help the user, who prefers to stick as closely as possible to the printed version.

CRV terminology. For computer-readable catalogues Albrecht (1982) has introduced the terminology of 'header' and 'trailer'. The 'header' is the description of the content of the catalogue. It comprises what we called 'data explanation', 'purpose' and 'location in space and time', and the explanation of the column headings. The 'trailer' provides the 'notes and remarks' and the 'bibliographic references' of the printed version. I shall come back in Chapter 9 to some of the problems raised in this chapter.

Notes on Chapter 5

This chapter is largely based on a paper by Jaschek (1984). The author is aware that he emphasizes – again – the need for careful descriptions, as has been done already in previous chapters. The excuse for such repetition is simply that this is often not done in the catalogues, as any user can see. It is rather frustrating to have to guess what an author did!

References

Albrecht, R. (1982) In *IAU Coll. 64, Automated data retrieval in astronomy*, Jaschek, C. and Heintz, W. (ed.), Reidel, p. 87

Argelander, R. (1859–1862) *Bonner Sternwarte Beobachtungen* 3, 4 and 5

Fresneau, A. (1981) *IBCDS* 20, 110

Heckmann, O. and Dieckvoss, W. (1975) *AGK3*, Hamburg-Bergedorf

Hoffleit, K. and Jaschek, C. (1982) *The bright star catalogue, fourth edition*, Yale University Observatory

Jaschek, C., Conde, H. and Sierra, A. C. (1964) *Publ. Ser. Astr. La Plata* 28, part 2

Jenkins, L. F. (1963) *General catalog of trigonometric stellar parallaxes*, Yale University Observatory

Mermilliod, J. C. (1988) *AA Suppl.* (in press)

Ochsenbein, F. (1982) PhD thesis, University of Strasbourg

Wilson, R. E. (1953) *General catalog of stellar velocities*, Carnegie Institution of Washington, Publ. No. 601

6

Designation of astronomical objects

In this chapter I shall discuss the designation problems alluded to in Chapter 4, because these problems turn out to be of fundamental importance for many of the following chapters.[1]

6.1   Designation of stars

Everybody knows that the very bright star Sirius is also called α Canis Majoris because it happens to be the brightest star of this constellation. If one looks up the *Bright Star Catalogue* (*Fourth Edition*) (Hoffleit and Jaschek 1982) one finds the proper names of about 850 stars but, although poets, historians and amateurs may use them, very few professional astronomers use more than two dozen or so of these names. How was this system introduced? Ptolemy described the stars in his catalogue by their position in the constellation – thus 'the brightest reddish star in the mouth, i.e. the dog star, Sirius' (α CMa).[2]

Later, Islamic astronomers translated Greek star descriptions into Arabic, and when these came back to the Western world, the Arabic names were adopted, in many instances as 'new' names. Thus, Mintaka, the belt (δ Ori), came from Greek 'the preceding one of the three belt stars'. The transmission of the Arab names came through Spain, both through translations and the *Alphonsine Tables* of 1252. This latter work was reprinted a number of times between 1483 and 1641 and constitutes one of the largest sources of Arabic star names.

A specific designation for stars was introduced by Piccolomini in 1539; he used capital Latin letters. His scheme was not widely accepted. Then, in 1603, Bayer published an Atlas (Uranometria) in which he denoted stars by Greek letters. These letters were assigned both in order of brightness (α brighter than β) and in order of position in the constellation. Despite the fact that this is a mixed procedure, the Bayer designations have essentially

Table 6.1. *The Durchmusterungen*

Area covered	Abbreviation	Declination zone	Number of stars	Approx. limiting magnitude
Northern	BD	+90° to −2°	324 188	9.5
Southern	SBD	−2° to −23°	133 659	9.9
Córdoba	CD	−23° to −90°	613 953	10
Cape	CPD	−18° to −90°	454 875	9.2 (phot.)

been kept up to now. The only permanent addition to the system has been the star numbering introduced in the *Historia Coelestis Britannicae* by Flamsteed (1725); this number was attributed according to the right ascension at the epoch. It is a usable system, provided that the constellation boundaries are carefully defined – which, in fact, they were not, neither in Bayer's nor Flamsteed's time. The official delineation of all accepted constellations was made by the International Astronomical Union in 1932, based upon work by Delporte (1930).

When in the nineteenth century Argelander (1859–62) with his collaborator Schönfeld (1852–66) and Thome (1892) did the *Durchmusterungen* – called respectively the northerly (BD) the southerly (SBD) and the Cordoba Durchmusterung (CD) (see Table 6.1) – a new system was introduced.

For example, α CMa is called SBD −16°1591, denoting that Sirius was (at equinox 1855.0) the 1591st star in the zone −16°. (The zone −16° extends from −16°0′0″ to −16°59′59″, equinox 1855.0.)

This numbering system tells us immediately whether a star is visible at a given latitude but it does not indicate its right ascension or its approximate position within the zone.[3]

A complication arose immediately when the Cape Observatory produced the Cape Photographic Durchmusterung (CPD), which covered the same zones as Córdoba. To distinguish between them, one must specify CD −50°126 or CPD −50°126 (for example) because −50°126 refers to a different star in each catalogue.

This system of designation is certainly useful for most of the brighter stars down to visual magnitude $V \simeq 8$, at least in the north, or to $V \simeq 9$ in the south, or photographic magnitude $m_{pg} \simeq 10$.

Obviously, if one is interested only in bright stars, one could use exclusively the BD, SBD, CD or CPD designation for all stars, with a

convention to suppress ambiguities between CD and CPD, plus a convention for the equinoxes. (Whereas BD and SBD have equinox 1855, CPD
and CD have chosen 1875.) The usual convention is to use BD between
+90° and −2°, SBD between −2° and −22°, CD between −22° and −52°
and CPD numbers between −52° and −90°.

Next came the designation introduced in the *Henry Draper Catalog* by
A. Cannon, the so called HD number. The HD numbers order stars by
order of increasing right ascension (equinox 1900.0) and decreasing declination. So, for example,

> HD 48910 has coordinates RA = 6 h 40 m 6 s, dec. = −78°75′,
> HD 48911 has coordinates RA = 6 h 40 m 7 s, dec. = +60°36′,
> HD 48912 has coordinates RA = 6 h 40 m 7 s, dec. = +30°23′,
>
> . . .   .
>
> HD 48915 has coordinates RA = 6 h 40 m 7 s, dec. = −16°35′.

Thus, Sirius = α CMa = BD −16°1591 = HD 48915.

Harvard photometrists (Pickering, 1908) had previously created a
catalogue of bright stars (BS) for which they obtained revised photometry.
(Harvard Revised Photometry = HR.) All stars with V ⩽ 6.5 have a BS =
HR number, which adds another designation. Thus, Sirius = α CMa = BD
16°1591 = HR 2491 = HD 48915. Sirius, of course, is also a visual double
star and, as such, it carries the designation ADS where ADS stands for
'Aitken Double Stars' (Aitken 1932). Sirius also has a radial velocity and
this is given in the *General catalog of stellar radial velocities* (Wilson 1952),
where Sirius carries the designation GCRV.

One could, of course, continue this listing of different parameters, but
the principle seems clear: each type of observation leads to a different
designation. Generally, in catalogues, the stars are ordered by right
ascension as in the HD catalogue but, unfortunately, at different
equinoxes; the GCRV uses 1950.0, for instance.

Variable stars are named by means of a different system: they are named
by constellation and within constellations by order of discovery but,
instead of starting with V1 Aquila, they start with R Aql, S Aql . . . Z Aql.
Then one uses RR, RS and so forth until ZZ. Next comes AA, AB . . .
(except J) up to QZ and, only afterwards, does good sense prevail so that
variables are called V335 Aql . . . etc. There are some exceptions: very
bright stars are designated by their habitually used names, such as β Ori, P
Cyg and l Car.

6.2   Designation of objects in general

It seems pointless to go over all the systems used to designate astronomical objects but let me simply state that objects may be designated by:

- a name (Sirius)
- a constellation (α CMa)
- a catalogue number based on some observational order, e.g. by declination ($-16°1591$), or by right ascension (HD 48915)
- a catalogue number based upon a temporal order (V431 Sco)
- a reference to a physical characteristic (ADS 5423)
- a reference to an instrument (3C 129, i.e. the 129th object of the third Cambridge survey)
- a reference to a wavelength range (Sco X-1, i.e. the first X-ray emitter in Sco)
- a reference to its nature (PSR 0031-07, i.e. the pulsar located at 00 h 31 m $-07°$ (1950.0))
- a reference to an observatory (Lick H alpha 52, i.e. the 52nd object discovered on an H alpha survey carried out at Lick observatory)
- a reference to a field in which the star is contained (SA 16-106, i.e. star 106 of Selected Area 16)
- a reference to another object in which the object lies (NGC 125-7, i.e. star 7 of NGC 125)
- an astronomer's name (Przybilski's star)
- an individual name (Rosette nebulae)

and very probably some more possibilities that I am unaware of. The system is certainly colourful and would even be usable if astronomers cared a little bit about designations. If somebody calls an object W3, he should specify whether W stands for Westerhout, in which case it designates a radio source, Wray – it is then an emission line star – or Walker, Wilson or Woolley – in which three cases it is a star.

When no comprehensive list of abbreviations existed, some ambiguity was unavoidable. Fortunately, Fernandez *et al.* (1983) compiled the *First dictionary of nomenclature of celestial objects*, which provides 34 pages of catalogue abbreviations plus notes and bibliographic references. This is an extremely valuable dictionary, which is the standard reference. A supplement to this dictionary has appeared recently (Lortet and Spite, 1986).

To improve things gradually it seems best to apply a few practical rules on designations, which are but common sense (Jaschek 1986):

(a) make sure that the designation you use is explained. Do not write −36°2163 but add whether it is CD or CPD and use the abbreviation recommended in the 'Dictionary';

(b) always use two identifiers per object, either one name plus a position, or two names. The system of the positions must be specified too. Do not write 32°42′ but l = 32°42′;

(c) if you describe objects that are unnamed, provide yourself a clear designation for the object. Do not call it object 1 but ZIL 1 (object No. 1 discovered by J. Zilch) and check first in Fernandez *et al.* (1983) that ZIL has not been used already for ZILINSKY. This is important, because if you do not name your object, somebody else will do it and he will call it MAO 1 (Mount Apple Obervatory) or ZI 1 (Zilch 1) or Z1, with much ensuing confusion, for which you are principally responsible.

6.3   Dictionary of synonyms

By now the reader will probably be wondering whether anyone has had the idea of composing a dictionary of synonyms, i.e. a dictionary in which all the different names of an object are given. One could then code the dictionary in such a way as to enter any name (among those that occur in the dictionary) and get all synonyms.

Such a 'dictionary' was effectively started by Jung (1971); it is described by Ochsenbein *et al.* (1977). A more recent paper by Ochsenbein (1985) provides an up-dated summary of the content of the dictionary, which in 1986 contained roughly:

600 000 stars
60 000 galaxies
6000 other objects (outside the solar system).

For each object, up to 35 designations are given, drawn from about 300 different lists and catalogues. This 'dictionary' is currently the biggest enterprise of this type. We shall come back to its organization in Chapter 10. Its historical name is *Catalogue of Stellar Identifications* (CSI) but the inclusion of non-stellar objects implies that it is a 'catalogue of astronomical object identifications' (CAOI).

6.4   Designations and positions

A little reflection on the designation problem has led many people to look for a universal solution. Such a miraculous solution is rediscovered regularly: it consists of dropping the names and keeping only positions as identifiers. Let us consider the implications in two parts; we shall deal first with the stellar problem and secondly with the non-stellar problem.

For stars, the proposal has been discussed a number of times. Advocates of it insist that one object has only *one* position instead of thirty names (or more). This means that we replace the string:

$$\alpha \, \text{CMa} = \text{Sirius} = \text{BD} -16°1591 = \text{HD } 48915 = \text{HR } 2491$$

by

$$\text{RA}(1950) = 6 \, \text{h} \, 42 \, \text{m} \, 56.7 \, \text{s} \quad \text{dec.}(1950) = -16°38'51''$$

Since equatorial coordinates change with time because of precession, one must either adopt one equinox by convention (for instance 1950.0 or 2000.0) or use a system insensitive to precession, such as the galactic coordinate system. For the latter possibility see Eichhorn (1977).

Now, the choice of Sirius as an example shows immediately the difficulties with the proposal, because, first of all, Sirius is a visual binary. Each of the components changes position with time and, in order to distinguish between components one must add the component designation – α CMa A or α CMa B.

Secondly, the position of Sirius changes with time, not only because of precession, but also because of its proper motion ($\mu = 1.38$ arc sec/year). The position of any star given for a certain epoch becomes useless after some time because of proper motion but, unfortunately, one cannot predict in advance what the proper motion of a star will be, without measuring it.

Thirdly, one should not forget that each century has produced an improved position of Sirius, and the position given above will be subject to similar improvement. Thus, either one changes the position as soon as a better position becomes available (but then we need a dictionary to discover what the object was called before), *or* one does not change it if a better position becomes available, and then one needs a dictionary the other way around. So, for stars, the system does not seem very useful – and we already have the CAOI.

Let us turn next to non-stellar objects. We notice first that the number of non-stellar objects catalogued at present is an order of magnitude (or so) smaller than the number of stars catalogued.

We observe next that non-stellar objects have widely varying angular diameters – compare for instance the Hyades ($\sim$7°), M31 (Andromeda galaxy) ($\sim$100 arc min), the planetary nebulae M1–4 ($\sim$4 arc sec), and the quasar 3C293 ($\sim$0.002 arc sec).

It is thus clear that the precision with which the position of an object may be given, cannot be the same for all objects. For the quasar, ± 1 arc sec is fine but, for M31, the photocentre cannot be determined to better than ± 1 arc min.

To circumvent this difficulty, proponents of identification by coordinates advocate the use of a position plus an object identifier. So, the planetary nebula NGC 6881 can be designated

PN 74 + 21°1

where PN tells us that the object is a planetary nebula, and $+74°$ and $+21°$ give the position in galactic coordinates. The last digit is an identifier within the 'box', if there is more than one object in it, in order of discovery. Therefore, in order to designate an object we must know its nature – be it a quasar, a planetary nebula or a galaxy – plus the coordinates. Observe that, in one example quoted, we cannot even locate the object in the finder of a telescope, because the field of a finder is usually smaller than 1°. On the other hand, notice also that:

(a) to designate 2000 planetary nebulae we need a type of object (PN) plus five figures and one sign – eight characters for 2000 objects! If there is more than one object with the same five figures and sign, a ninth character has to be added;

(b) we need to be sure that it is a planetary nebula, a fact which, according to Acker (1986), can be assured for only two thirds of the objects listed as PNs.

Thus it seems that neither in the stellar nor in the non-stellar case is designation by coordinates alone really practicable.

The source of the difficulty can best be seen by writing:

designation = name,
        position = parameter.

I think that readers will agree that it is easier to designate the author of this book by the name 'Carlos Jaschek' than to characterize him by

AC 00.55.1 +45.39 (1984, July)

AC = Argentine Citizen; coordinates of my position on Earth on 1984 July (I was then at Trieste). I hope my friends can still recognize me even if I go back to Strasbourg and am designated by

AC 0.0.78 +48.35 (1986, March).

In my opinion, we should continue attributing arbitrary designations to objects as the means of identifying them and consider their positions and physical nature as additional information, *not to be put in the name*.

As I said before, this is my personal opinion, which is not shared by all other colleagues. For a different point of view see, for instance, Dickel *et al*. (1987).[4]

We shall describe next the designation practices accepted by the IAU. Former IAU resolutions concerning designations were summarized by Jaschek and Pecker (1979). A forthcoming publication by Mead, ordered by types of objects, will provide more details, and the forthcoming IAU style book (1988 edition) will give a summary of current practices.

6.5   Officially accepted designation practices

Table 6.2 lists the official practices. In the explanations, 'see text' refers to matters already dealt with in this chapter; H and M stand for hours and minutes of right ascension and m for fractions of a minute; D and M designate degrees and minutes of declination and + the sign for declination; LLL means degrees of longitude and BB degrees of latitude.

A recent IAU resolution (Swings, 1986) provides some additional clarification of the status of designation practices and is here reproduced in full:

> *Resolution C3: Astronomical Designations*
> Commission 5, with a view to avoiding confusion,
> recommends strongly   That IAU resolutions which concern the designation of astronomical objects outside the solar system be forwarded to the Working Group of Commission 5 on Designations for its advice before being passed on to the General Assembly.
> Furthermore, Commission 5,
> recognizing   the many benefits that would follow from the clear and unambiguous identification of all astronomical journals and other sources of data,
> strongly urges   (1) that all astronomers follow the IAU recommendations on the designation of objects outside the solar system that were adopted by Commission 5 in 1979 (IAU Transactions XVIIB, 87–88) and the

Table 6.2. *Officially accepted methods of designation*

Object type	Designation
Point sources	
Brighter stars	BS = HR
	HD
	Durchmusterung number according to convention. See text.
Visual binaries	IDS (*Index Catalogue of visual Double Stars*) HHMM ± DDMMX where X = A, B . . . designates the components. Equinox 1900.
Variable stars	Designation as given in the *General catalogue of variable Stars*. See text.
	Equinox 1900.
	For stars in clusters or galaxies one uses V 5-NGC 5272, meaning the fifth variable discovered in this globular cluster.
Extended sources	
Star clusters in our galaxy	C for cluster followed by 1950.0 equatorial coordinates HHMM ± DD.d, d = decimal of declination degree.
	Not to be changed if cluster position is improved.
Stellar assoc. in our galaxy	No consistent system exists.
Planetary nebulae:	
– in our galaxy	PN – LLL + BBN, N = number of discovery in field $1° \times 1°$, PN = planetary nebula.
– in other galaxies	Galaxy name – PN – x, x = current number in order of discovery.
Supernova	By order of events: SN 1985 A, SN 1985 B . . . SN 1985 Z, followed by aa, ab . . . , bb, . . . bz.
Radio sources	Can be designated by
	(a) name of the survey plus a sequential number like 3C123, provided the designation uses the abbreviations proposed by Fernandez *et al.* (1983, 1986);
	(b) Source name plus coordinates, i.e. source name X HHMMSS.SS + DDMMSS.S;
	X stands for the B or J system. If galactic coordinates are used X = G.
Pulsars	Same as galaxies, preceded by PSR, instead of source name.
Molecular clouds	Tau MC (molecular cloud in Taurus) LLL.L + BB.B.
X-ray sources	Name of satellite, number of catalogue, HHMM.m + DD.MM.
Galaxies	Name or name plus coordinates (1950.0) (HHMM.N + DD.D);
	NGC 55 or A 1955 + 40, (A = anonymous).

supplementary precepts that are given below in the Memorandum on Designations adopted in 1985,

(2) that the editors of astronomical journals draw the attention of authors to these recommendations, preferably by providing a summary, and request referees to refer back any papers or tabulations that do not provide satisfactory designations, and

(3) that the Space Telescope Science Institute adopt these recommendations for objects discovered with the Space Telescope.

Memorandum on Designations – New Delhi 1985

(1) The IAU Style Book, 1986 Revision, will provide rules on designations to be used for constellations, stars and other astronomical objects.

(2) Astronomers should consult *The First Dictionary of the Nomenclature of Celestial Objects* by Fernandez, Lortet and Spite and its supplement (1983, *A. & A. Suppl.* **52**, No. 4, and 1986 *ibid.* **64**, 329) for the designations of various types of astronomical objects already in the literature and to avoid duplication when proposing designations of new objects.

(3) The following precepts are now added:

(i) The IAU-approved three-letter abbreviations for the constellations together with LMC and SMC for the Magellanic Clouds should be used. These abbreviations should not occur with any other meanings.

(ii) Abbreviations of abbreviations (e.g. 'N' for 'NGC') should never be used.

(iii) Personal names such as 'Gum Nebula' should be preserved as in the First Dictionary.

(iv) New acronyms for abbreviating catalogues, types of objects, authors' names, observatories, etc. should have at least two letters.

(v) The list of types of objects (e.g. GCL, SNR, etc.) given in the First Dictionary should be followed closely.

(vi) Specific references are needed for acronyms that do not appear in the First Dictionary and for those that appear in the First Dictionary with classification E (explain) or those with S (systematic) which could be ambiguous (e.g. OH).

(vii) For designations based on coordinates
- use truncated coordinates, not rounded up ones;
- use explicit leading zeroes and the declination sign;
- use decimal points if appropriate;
- adopt the EINSTEIN extended format if possible
  (e.g. acronym HHMMSS + DDMMSS
  or acronym HHMMSS.SS etc. +DDMMSS.S etc.);
- when necessary to distinguish old names based on Julian 2000, precede the right ascension with a J in the latter case; (note: Relevant IAU resolutions concerning J 2000 and the new astronomical constants are summarized in United States Naval Observatory Circ. 183);
- adopt for galactic coordinates the prefix G
  (e.g. acronym GLLL.LL +BB.BB); and
- if a coordinate designation includes the catalogue name, rather than the type of object, do not change the designation when the coordinates are improved.

(viii) The recommended form for designation of individual objects inside a larger object is
(e.g. LARGE: Acronym Number).

(ix) When objects are designated on finding charts, the coordinate equinox, the scale, and the N–S and E–W orientations should be indicated clearly.

For extended objects the situation is still in a state of flux. The proceedings of a recent discussion on the subject (Dickel *et al.* (1987)) show very well the complications to be overcome, although some of the solutions proposed do not seem to be very practicable. In conclusion, it may be said that we are still a long way from a consistent designation system, despite the progress made.

Past experience leaves doubts about the usefulness of IAU resolutions. As an example, (Jaschek and Pecker, 1979) the abbreviation of constellation names by three letters was the subject of IAU resolutions at the General Assemblies of 1922, 1952, 1955 and 1970 – and 1985. It took sixty years to introduce a standard practice for constellations and there are only 88 constellations!

A possible remedy is to put pressure on the editors of astronomical journals to apply effectively IAU recommendations in all papers they accept: it is probably easier to convince the editors of a hundred journals, rather than twenty thousand astronomers. Editors have been asked to apply such measures but the outcome has not been very encouraging, in the sense that no journal has applied strict measures, or tried to enforce a consistent policy. This probably means that not even the editors have a clear feeling that this is an important problem.

6.6 Designations and priority rights

Society stimulates inventors by according patents, which lead to economic rewards. In pure sciences, the stimulus consists of awarding a priority right on the discovery made. No economic reward is directly associated with a discovery, although promotions, prizes or other honours may be bestowed upon the discoverer. This explains why priorities are so highly regarded and why so many disputes have started over it, disputes that may rage over years and may even involve nationalistic aspects, as in, for instance, the circumstances of the discovery of planet Neptune.[5]

If a discovery is considered important, the name of the discoverer is attached to it: we all use Kepler's laws, Newton's principles, the Mossbauer effect and Pauli's principle. Since the number of great discoveries is small and the names of individuals one can retain in memory is also limited, it is clear that only a few scientists can be honoured this way.

In natural sciences such as botany, zoology and mineralogy, scientists have used new discoveries to honour their peers, by either attaching the discoverer's name to a mineral or species, as in Baugnerite (a mineral), or by honouring scientists in a name, as in Fuchsia (a flower) or Gerardia (a plant). Since the number of plant, animal or rock species is large, such a system can honour a large number of scientists.

Such a practice can also be found in astronomy, more specifically in the astronomy of the solar system. One finds lunar craters named after outstanding scientists, comets named after their discoverers and asteroids designated with names selected by their discoverers. All these names are made official through the appropriate Commissions of the International Astronomical Union.

In non-solar system astronomy, such practices are less frequent. There are a few objects associated with scientists – the Herbig-Haro objects, Stephan's Quintet and so on. Spite and Lahmek (1982) and Fernandez *et al.* (1983) have provided lists of designations of this type, which contain less than one hundred objects but, in general, such designations are an exception rather than the rule. Unless they are officially recognized, the 'personalized' names tend to be replaced in course of time by more neutral designations.

There is, however, another practice, which is more common – that of incorporating the name of an astronomer into the designation of a catalogue with which he is associated. There is a 'Henry Draper' number (HD), although the spectral classifications given in that catalogue were not made by him, the Aitken Double Star Catalog (ADS), to which Aitken made an important contribution, and the Messier (M) numbers from a catalogue of objects observed by Messier. Such 'personalized' names are, however, not the rule: we find the *Bonner Durchmusterung* (BD) by Argelander and the *New General Catalogue* (NGC) by Dreyer, with designations for clusters and galaxies, which are not associated with the names of their authors.

Since any catalogue refers to only some of the parameters of an object, it is clear that the same object may feature in several catalogues. Thus, stars having large proper motions were detected and catalogued by Luyten, although the stars may have been listed already by someone else. Then, a spectroscopist may observe those large proper motion stars and discover among them a number of white dwarfs. The list of these degenerate stars is published as a new discovery, although the star was only observed because somebody else had discovered its large proper motion.[6]

Such practices of re-discovery are acceptable as long as proper credit is given to the author of the data in the first list. If this is not done, we are in danger of a possible infringement of professional ethics (see, for instance, Merton 1973). There are numerous reasons for infringement, ranging from voluntary omission, to forgetfulness, editorial policies or historical reasons. The last reason may arise because it is impossible to retrace the whole history of a subject, so that individual contributions when (say) thirty years old slide into the background of 'generally accepted' facts or data.

Since the motives an author had for not quoting the work of a colleague are never fully known, long polemics may result. The original author feels deprived of his priority rights and reacts more or less polemically according to his or her character.

Notes on Chapter 6

1. For the classic star designations see, for instance, Lacchini (1959). Arab star names are discussed, for instance, by Samaha (1937), Mensard (1965) and in the classic book by Allen (1963). The latter is a reprint of the original 1899 version and should be used with some care. The book by Kunitzsch (1959) is based upon later research and should be preferred. Some useful comments on star names can also be found in Heuter (1986).

2. On the reddish colour of Sirius a large literature has developed. The classic studies are by Boll (1916) and See (1927). Both papers are usually omitted in recent papers on the subject.

3. The history of the *Durchmusterungen* can be read in Hermann (1984). Designation problems have been discussed a number of times, for instance, in Collins (1977), Spite (1977), Lortet and Spite (1979), Jaschek and Pecker (1979), Murray *et al.* (1979), Ochsenbein and Spite (1982), Fernandez *et al.* (1982), Polechova (1982), Ochsenbein and Bischoff (1982), Jaschek (1985).

4. The problem of substituting designations by positions was discussed, for instance, in 'The construction and use of star catalogues' conference. See specially Herget (1967). A recent proposal of this type is that of Martinez (1987).

5. Priority rights and related questions are discussed, for instance, in Merton (1973), Ziman (1984) and in other books on the sociology of science.

6. For a polemical paper on the subject see Luyten (1986).

7. A recent development has been the creation of a 'clearing house' for advice on designations. At the time of writing, five astronomers are acting as advisors: M. C. Lortet, Observatoire, 92190, Meuden, France; J. Mead and W. Warren, Goddard Space Flight Center; C. Jaschek, CDS; H. Dickel, Astronomy Dept., University of Illinois at Urbana-Champaign, 1011 W. Springfield Ave., Urbana, IL 61801, USA. The addresses of the Goddard Space Flight Center and of the CDS are given on page 133.

References

Acker, A. (1986) *Les nebuleuses planetaires*, Comptes Rendues Journées de Strasbourg, Observatoire de Strasbourg, p. 1

Aitken, G. R. (1932) *New general catalog of double stars within 120° of the Pole*, Carnegie Institution of Washington, Washington DC

Allen, R. H. (1963) *Star names: their lore and meaning*, Dover Publ.

Argelander, F. (1859–62) *Bonner Stern-Verzeichnis Sektion 1–3, Bonner Sternw. Beobachtungen* **3**, **4** and **5**

Bayer, J. (1603) *Uranometria*, Augsburg

Boll, F. (1916) *Antike Beobachtungen farbiger Sterne, Abh. Koeniglicher Bayerischen Akad. Wissensch. Phil. Hist. Kl.* **30**, Abh. 1

Cannon, A. J. and Pickering, S. (1924) *The Henry Draper catalog*, Harvard Annals Vol. 91–100, from 1918 to 1924

Collins, M. J. (1977) In *IAU Coll. 35, Compilation, critical evaluation and distribution of stellar data*, Jaschek, C. and Wilkins, G. A. (ed.), Reidel, p. 25

Delporte, E. (1930) *Atlas Celeste*, Cambridge University Press

Dickel, H. R., Lortet, M. C. and de Boer, K. S. (1987) *AA Suppl.* **68**, 75

Eichhorn, H. (1977) In *IAU Coll. 35, Compilation, critical evaluation and distribution of stellar data*, Jaschek, C. and Wilkins, G. A. (ed.), Reidel

Fernandez, A., Lortet, M. C. and Spite, F. (1982) In *IAU Coll. 64, Automated data retrieval in astronomy*, Jaschek, C. and Heintz, W. (ed.), Reidel, p. 203

Fernandez, A., Lortet, M. C. and Spite, F. (1983) *AA Suppl.* **52**, No. 4

Flamsteed, J. (1725) *Historia coelestis Britannicae*, London

Gill, D. and Kapteyn, J. C. (1896–1900) *Cape photographic Durchmusterung*, Cape Ann. **3**, (1896) **4**, (1897) and **5**, (1900)

Herget, P. (1967) *AJ* **72**, 587

Hermann, D. B. (1984) *The history of astronomy from Herschel to Hertzsprung*, Cambridge University Press

Heuter, G. (1986) *Vistas in astronomy* **29**, 237

Hoffleit, D. and Jaschek, C. (1982) *The bright star catalog, fourth edition*, Yale University Publ.

Jaschek, C. and Pecker, J. C. (1979) *IBCDS* **16**, 57

Jaschek, C. (1985) In *Data handling in astronomy and astrophysics*, Hauck, B. and Sedmak, G. (ed.) *Mem. Soc. Astron. Italiana* **56**, Nos. 2 and 3

Jaschek, C. (1986) *QJRAS* **27**, 60

Jung, J. (1971) *IBCDS* **1**, 3

Kunitzsch, P. (1959) *Arabische Sternnamen in Europa*, Wiesbaden

Lacchini, G. B. (1959) *Contr. Trieste*, No. 236

Lortet, M. C. and Spite, F. (1979) *IBCDS* **16**, 55

Lortet, M. C. and Spite, F. (1986) *AA Suppl.* **64**, 329

Luyten, W. J. (1986) *Proper motion surveys, LXVII*, University of Minnesota

Martinez, A. (1987) *IBCDS* **33**, 69

Mensard, H. (1965) *Ann. Toulouse* **17**, 63

Merton, R. K. (1973) *The sociology of science*, University of Chicago Press

Murray, C. A., Gliese, W. and Hoffleit, D. (1979) *IBCDS* **16**, 71

Ochsenbein, F., Egret, K. and Bischoff, M. (1977) In *IAU Coll. 35, Compilation, critical evaluation and distribution of stellar data*, Jaschek, C. and Wilkins, A. G. (ed.), Reidel, p. 31

Ochsenbein, F. and Bischoff, M. (1982) In *IAU Coll. 64, Automated data retrieval in astronomy*, Jaschek, C. and Heintz, W. (ed.), Reidel, p. 211

Ochsenbein, F. and Spite, F. (1982) In *IAU Coll. 64, Automated data retrieval in astronomy*, Jaschek, C. and Heintz, W. (ed.), Reidel, p. 199

Ochsenbein, F. (1985) *IBCDS* **28**, 131

Piccolomini, A. (1539) *Della sfera del mondo*, Venezia

Pickering, E. C. (1908) *Harvard revised photometry, Harvard Ann.* **50**

Polechova, P. (1982) In *IAU Coll. 64, Automated data retrieval in astronomy*, Jaschek, C. and Heintz, W. (ed.), Reidel, p. 207

Samaha, A. H. M. (1937) *Medd. Lund, II ser.*, No. 11

Schönfeld, E. (1866) 'Bonner Sternverzeichnis'. Sektion 4, in *Bonner Sternw. Beobachtungen* **8**

See, J. (1927) *AN* **229**, 245

Spite, F. (1977) In *IAU Coll. 35, Compilation, critical evaluation and distribution of stellar data*, Jaschek, C. and Wilkins, G. A. (ed.), Reidel, p. 25

Spite, F. and Lahmek, B. (1982) *IBCDS* **22**, 105

Swings, J. P. (1986) *Transact. IAU XIX B*, Reidel, p. 40.

Thome, J. M. (1892–1932) Córdoba Durchmusterung, Resultados Obs. Córdoba **13, 14, 16, 17, 18**, and **21** (1932)

Wilson, R. F. (1952) *General catalog of stellar radial velocities*, Carnegie Inst. Washington Publ. No. 601

Ziman, J. (1984) *An introduction to science studies*, Cambridge University Press

7

Catalogues

We have dealt in Chapter 3 with data in astronomy and in Chapter 5 in some detail with catalogues, which are one of the classical ways of presenting data. We shall discuss in this chapter the important question of how one may find the catalogues one is interested in.

7.1 Surveys of catalogues

Since most of the oldest catalogues are astrometric, it is not surprising to find that the best surveyed area is that of astrometric catalogues, of both the observational and compilation types. (For the terminology the reader is referred to Chapter 5.)

Let us start with the main publications dealing with catalogues. The first modern works are Ristenpart (1901) and Kobold (1926), which summarize very well the classic literature and are still readable for this reason. From the thirties we find the discussions of astronomical catalogues by von der Pahlen (1937) and Bok (1937). Since both authors were concerned with galactic astronomy, they wrote careful discussions, still worth reading, on catalogues of photometry, radial velocities, proper motion and parallaxes. Eichhorn (1974) presented in his book, *Astronomy of star positions*, the most comprehensive discussion available of astrometric catalogues; the other data are not treated systematically.

Sevarlic *et al.* (1978) produced a comprehensive bibliography of astronomical catalogues – they list more than two thousand. In a second part they list photographic catalogues (Sevarlic *et al.* 1982).

Collins (1977) published a list of all catalogues printed in the period 1950–75. This is a very useful book, because of its comprehensiveness; its drawback is that it is not easy to find out which are the most important catalogues.

For initial orientation, the book by Seal and Martin (1982) is also useful;

data collections are among the many sources quoted in it. This work covers the period 1970–79.

The most recent systematic presentation of catalogues is the one by Jaschek (1984a), who lists the largest compilation catalogues in each field in machine-readable form and available on tape. In what follows we present an update of that paper. The survey is limited to the most recent compilation catalogues in each field. The computer-readable versions are obtainable from data centres (see Chapter 10).

Catalogues are discussed in order, grouped by object type and parameters. Besides each catalogue title are given the author's name, the date of publication and a second number in parentheses, which is the reference number to the computer-readable version (for details see Chapter 10). The number of objects in each catalogue has been rounded off and is given for orientation purposes only. We have not included small observational catalogues.

7.2 Data on stars

7.2.1 *Positions and proper motions*

The most widely-used catalogue is the SAO (Smithsonian Astrophysical Observatory) by Haramundanis (1966) (1001). It provides positions and proper motions for 259 000 stars on the whole sky. However, it should be noted that the SAO is not complete down to a given magnitude since, in crowded areas, some stars were omitted and that it is rather poor for $\delta < -60°$.

For the northern sky, the AGK3 by Heckmann and Dieckvoss (1975) provides data for 193 000 stars north of $\delta = -2.5°$. It comes in three versions (1002, 1061 and 1969), each differing in some details. For a discussion of these catalogues, see Fresneau (1981). No similar coverage exists for the southern sky.

Positions of about 3.6×10^5 stars are known to an arc second but most of them are brighter than tenth magnitude. Proper motions are known for about the same number of stars (Ochsenbein *et al.* 1981) but, for instance, in the AGK3 only one-third have 'significant' proper motions, i.e. larger than three times the mean error (Fresneau 1978). For stars with very large proper motions one can use Luyten (1979, 1980) (1098), which provides 58 000 stars with annual proper motions in excess of 0.2 arc sec per year.

Giclas, Burnham and Thomas (1978a, b) (1079 and 1112) provide another set of large proper motions in both hemispheres, down to $-30°$.

For general purposes, there is also the Cape Photographic Durmuste-
rung by Gill and Kapteyn (1895–1900) (1108), which provides data for
455 000 stars, and the Córdoba Durchmusterung by Thome (1892–1932)
(1114), which provides data for 614 000 stars. The first covers the region
$-18° < \delta < -89°$, the second $-22° < \delta < -89°$. The northern part of the
sky (Bonner Durchmusterung (Argelander 1859–62) and Bonner Südliche
Durchmusterung (Schönfeld 1886)) is available under catalogue numbers
1119 and 1122. All Durchmusterungen provide only approximate positions
and magnitudes, which should not be used unless nothing else is available.
A list of all recent meridian circle catalogues will be published in 1988 by
the CDS.

A general file on kinematic data of HD and HDE stars is given by
Mennessier *et al.* (1978). This catalogue provides what are considered to
be the best available data and contains 178 000 stars, of which 8×10^4
possess good proper motions.

In 1988 the *Catalogue of Guide Stars for the Space Telescope* will become
available – it provides positions and magnitudes for about 3×10^7 stars
down to the thirteenth magnitude.

7.2.2   *Parallaxes*

For trigonometric parallaxes, one may use the Jenkins catalogue
(1963) (1081), which contains data for 7 300 stars. However, only a fraction
of them have accuracies better than 10% or so. The so called 'nearby stars'
are compiled by Gliese and Jahreiss (1979) (5001, 5035). Gliese (1983)
reports that about 700 stars have this level of accuracy ($\Delta M < \pm 0.3$).

A new general catalogue of trigonometric parallaxes by van Altena is
forthcoming.

7.2.3   *Radial velocities*

The standard catalogue in this field is Wilson (1953) (3021), which
gives average radial velocities for 15 000 stars. Evans (1967) (3047) provides additional average radial velocities for 7800 stars; however, his
survey is incomplete and it is recommended that users also employ the
OCC by Abt and Biggs (1972) (3004), which provides bibliographic
references up to 1970, and Barbier (1975) (3038), which updates to 1974.
The last catalogue is now superseded (see below). A summary of the data
in these four publications may be found in Ochsenbein *et al.* (1981) (5026).

Barbier-Brossat and Petit (1986) (3108) update the literature to 1980.

However, no catalogue of average radial velocities is available. Radial
velocities are known for about 27 000 stars.

7.2.4 *Photometry*

The largest body of non-photoelectric measurements is that of the 3.6×10^5 stars reduced to a homogeneous system by Ochsenbein (1974) (5026). These values should be used only if photoelectric values are unavailable.

Photoelectric measurements are known for about 1.02×10^5 stars (Mermilliod 1984). The forthcoming *General Catalogue of Photometric Data* by Hauck will provide references to the system in which each of these stars was observed. At present, only a first version of this catalogue by Magnenat (1976) (2039) is available. It provides references to the system in which 3.2×10^4 stars were observed. The situation with regard to photometry in different wavelengths is given in Table 7.1.

Table 7.1. *Stellar photometry*

System	Author	Year	Cat.	No. of stars	Notes
a) *Range $\lambda < 3000$ Å*					
TD1	Thompson *et al.*	1978	5038	31 000	OC
Celescope	Davis *et al.*	1973	2006	5700	OC
ANS	Wesselius *et al.*	1982	2097	3600	OC
b) *Range 3000 Å $< \lambda < 10\,000$ Å*					
UBV	Mermilliod & Nicolet	1977	2035	53 000	M
	Mermilliod	1983	2089	26 000	U
	Nicolet	1978	2051	59 000	A
	Mermilliod	1987	2122	87 300	M, A
uvby	Hauck & Mermilliod	1985	2107	40 000	M, A
Geneva	Rufener	1981	2072	14 600	A
U_cBV	Nicolet	1975	2027	7200	M, A
Vilnius	North	1984	2109	2600	A
DDO	McClure & Forrester	1981	2080	2200	A
RI	Jasniewicz	1982	2075	5700	M, A
UBVRI...N	Morel & Magnenat	1978	2007	4500	M, A
UBVRI	Lanz	1986	2116	8390	M, A
c) *Range $\lambda > 1 \mu m$*					
$\lambda > 1 \mu m$	Gezari *et al.*	1982	2098	3×10^4	BC
IRAS*	Joint IRAS Science Working Group	1984	2125	2.4×10^5	OC

* This catalogue also includes observation of non-stellar objects.
Note: For the passbands of the different photometric systems see Golay (1974). In the column 'Notes', M = measurements, A = averages, U =update of measurements.

7.2.5 *Spectroscopy*

For unidimensional spectral types, the most important collection of data is the *Henry Draper Catalog* by Cannon and Pickering (1924) (3001) covering 225 000 stars; with the Henry Draper Extension this figure rises to 350 000 stars.

Houk is currently reclassifying all HD stars by declination zones; the first volumes of this work are already available: vol. I ($-89°$ to $-53°$), Houk and Cowley (1975) (3031); vol. II ($-53°$ to $-40°$), Houk (1978) (3051); vol. III ($-40°$ to $-26°$), Houk (1982) (3080); these provide types for about 97 000 stars.

MK spectral classifications for stars were collected by Jaschek *et al.* (1964) and for 36 000 stars by Kennedy (1983) (3078).

Mercedes Jaschek (1978) (3042) has produced a catalogue of selected spectral types for 31 000 stars, basing her 'best' spectral types on the literature available.

Spectral classifications for 1900 stars based upon the ultraviolet region have been published by Cucchiaro *et al.* (1979) (3053) (TD1 experiment) and by Heck *et al.* (1984) (3083) for 230 stars (IUE experiment).

Stars with spectral peculiarities are listed by Jaschek and Egret (1984) (3081); the catalogue lists about 25 000 stars belonging to 53 different groups.

Stellar abundances for 1000 stars in the form of (Fe/H) ratios are given in a bibliographic catalogue by G. Cayrel *et al.* (1984) (3096).

A critical catalogue of stellar abundance studies was established by Köppen (1987) (3900).

For further details on spectroscopic computer-readable catalogues, see Jaschek (1984a).

7.2.6 *Spectrophotometry*

The two standard catalogues in this field are Breger (1976) (3048), with data for 940 stars, and Ardeberg and Virdefors (1980) (3069), with data for 380 stars. Jacoby *et al.* (1984) (3092) have published high resolution spectrophotometry for 160 stars and Gunn and Stryker (1983) an atlas for 175 stars (3088).

There are three important spectrophotometric catalogues for the ultraviolet region, Jamar *et al.* (1976) (3039), giving data for 1800 stars, and its complement, Macau-Hercot *et al.* (1978) (2086), giving data for 400 stars. The third is the IUE flux catalogue by Heck *et al.* (1984) (3083), giving data for 230 mostly normal stars. There are a number of other catalogues providing data for smaller numbers of stars.

7.2.7 *Stellar rotation*

The latest catalogue is from Uesugi and Fukuda (1982) (3063). It provides average values for 6500 stars.

7.2.8 *Diameters*

Fracassini *et al.* (1981) (2061) have produced a bibliographic catalogue of the existing determinations for 4300 stars.

7.2.9 *Masses*

Popper has edited a critical compilation catalogue (1980), but it is not available on tape. For additional information see the section on 'Binaries'.

7.2.10 *Polarization*

The only catalogue on this subject is Mathewson *et al.* (1978) (2034). It provides data for 7500 stars but no update is available. A list of standards has been provided by Breger and Hsu (1981).

7.2.11 *Magnetic fields*

Didelon (1983), lists magnetic field measurements for 800 stars. No tape version exists.

7.2.12 *Binaries*

The *Washington catalog of visual double stars* by Worley and Douglas (1984) (1107) deals with visual binaries giving data for 74 000 objects; the catalogue of individual measures is not yet published. A new catalogue of visual binary orbits became recently available (Worley and Heintz, 1983) (5039); it gives orbits for 850 systems. The orbit catalogue of Batten *et al.* (1978) (5040), dealing with spectroscopic binaries, provides almost 1000 orbits. This catalogue has been updated by Pedoussaut *et al.* (1986) (4016). Physical parameters derived from these data are given by Kraicheva *et al.* (1980) (5041) for over 900 binaries. Eclipsing binaries are included in the catalogue of orbits of close binaries by Svechnikov and Bessonova (1984) (5042), which gives data for about 300 pairs. Wood *et al.* (1980) (6023) provide a finding list for observers of interacting binaries.

7.2.13 *Variable stars*

For variable stars, one may use the third edition of the *General catalogue of variable stars* by Kukarkin *et al.* (1971) (2011), which provides data for 23 000 stars, and its supplement (suspected variables) by Kukarkin

(1981) (2079), which gives data for 15 000 additional stars. Part of the fourth edition of the catalogue is already available (Kukarkin *et al.* 1985) (2113), listing 18 200 stars.

A recent bibliography of variable stars by Huth and Wenzel (1986) (6035) provides 290 000 references for 28 000 variables; a supplement by the same authors (1986) (6038) adds information on 5000 variables.

For 'general' compilation catalogues of stars, providing much data, we have the *Bright star catalog, Fifth revised edition*, by Hoffleit and Warren (1986) (5050); it provides data for 9100 stars brighter than 6.5 mag.

7.3   Non-stellar objects – general

There are two catalogues of non-stellar objects, namely the one by Sulentic and Tifft (1974) of 9000 objects (7001) and the *Master list of non-stellar objects* by Dixon (1976) of 1.8×10^5 objects (7002).

7.4   Non-stellar objects in our Galaxy

We turn next our attention to non-stellar objects in our Galaxy.

7.4.1   *Clusters*

The standard catalogue of open clusters is by Lynga (1987) (7092), and provides data for 1200 objects. Globular clusters and associations are dealt with in the bibliographic catalogue of Ruprecht *et al.* (1983) (7044) (140 objects); associations have also been an object of study by the same authors (1983) (7031).

Mermilliod (1987) (2124) takes on the more difficult problem of data for stars in open clusters and provides UBV values and MK classifications for 8900 and 2300 stars, respectively, in 223 clusters. Another important catalogue is Mermilliod's (1979) (3055) catalogue of radial velocities in clusters, giving data for 800 stars in 78 open clusters. A new update is in preparation.

There is no general catalogue that enables us to separate stars belonging to open clusters from field stars. However, Humphreys *et al.* (1984) (7046) have taken a step in this direction by compiling a catalogue that lists 2300 objects in associations. Mermilliod (1985) (4017) cross-identified HD, HDE and DM identifications for 9500 stars in open clusters.

7.4.2   *Nebulae and globules*

Dark nebulae are listed by Lynds (1962) (and updates); she provides 1800 objects (7007). Lynds (1965) (7009) also lists 1000 bright nebulae. Reflexion nebulae are listed by van den Bergh (1966) (7021) (160 objects). A list of globules is available from Wesselius (1979) (7057); it comprises 800 objects. Planetary nebulae are catalogued by Acker *et al.* (1981) (7024); they list 1500 objects.

7.4.3 *Miscellaneous*

Supernova remnants are catalogued by Clark and Caswell (1976) (7014); there are about 120 objects.

For H II regions one still has to use an old list by Sharpless (1959) (7020) with 300 objects. The more recent catalogue of Marshalkova-Polechova (1974) does not exist on tape.

Pulsars are catalogued by Seiradakis (unpublished) (7008); he lists 150 objects.

Radio stars are compiled by Wendker (1984) (2112); he lists 3500 objects.

7.5 Non-stellar objects outside our Galaxy

7.5.1 *Galaxies*

Galaxies are catalogued by de Vaucouleurs and de Vaucouleurs (1964) (7016). A new (1976) edition of this *Reference catalog of bright galaxies* is available from the University of Texas Press. Another sample of bright galaxies is the *Revised Shapley Ames catalog of galaxies* (1200 galaxies) by Sandage and Tamann (1981) (7051). The *Merged catalogue of galaxies* by Kogoshvili (1983) (7090) provides a combination of different sources and lists 32 000 objects.

7.5.2 *Parameters of galaxies*

The polarization of extragalactic nebulae is compiled by Eichen- dorf and Reinhardt (1979) (7006). CO observations are reported by Verter (1985) (7064). Infrared magnitudes and HI widths were catalogued by Aaronson *et al.* (1982) (7075) for 300 nearby galaxies. A catalogue of 3300 galaxies south of $-30°$, observed spectroscopically, is given by Fairall *et al.* (1985) (7079).

A catalogue of redshifts is given by Rood (unpublished) (1980) (7036) for 4000 galaxies. Morphological types of galaxies are given by Vorontsov-Velyaminov and Arkhipova (1962–68) (7062) for 29 000 galaxies.

7.5.3 *Special types of galaxies*

A list of 1500 Markarian galaxies has been compiled (7061). The *Atlas of peculiar galaxies* by Arp (1966) (7074) contains 350 galaxies.

Seyfert galaxies are listed by Weedman (1978) (7047); he lists 120 objects. A list of galaxies having an ultraviolet excess (400 objects) is given by Mitchell *et al.* (1982) (7063). An optical catalogue of radiogalaxies is given by Burbidge and Crowne (1979) (7033); it contains 500 objects.

Isolated pairs of galaxies are catalogued by Karachentsev *et al.* (1972) (7077); 600 pairs are given, and isolated triplets of galaxies. Karachentsev (1986) (7083) provides a list of 84 triplets.

7.5.4 *Clusters of galaxies*
A catalogue of Abell and Zwicky clusters is given by Kalinkov *et al.* (1979) (7004) for 11 800 clusters. The Zwicky *Catalog of galaxies and clusters of galaxies* (1961–68) for 30 000 objects is available (7049). Redshifts for the nearby clusters have been compiled by Baiesi-Pillastrini *et al.* (1984) (7071) for 600 objects. Southern groups and clusters of galaxies are given by Duus and Newell (1977) (7028) for 10^3 objects. The *CEDAG catalogue of clusters of galaxies* by Fernandez *et al.* (1984) (7088) provides data for 10 500 clusters.

7.5.5 *Quasars*
The *Asiago catalogue of quasistellar objects* by Barbieri *et al.* (1983) (7069) lists 2000 objects. The *Catalogue of quasars and active galaxies* (Veron-Cetty and Veron 1987) (7093) provides 2900 objects.

7.5.6 *Magellanic clouds*
A catalogue by Florsch *et al.* is announced for 1988 for the Small Cloud.

7.5.7 *Radio and X-ray sources*
For radio sources there are four surveys on magnetic tape, namely:

- *Third Cambridge radio catalogue* by Bennett (1961) (7010) (300 sources);
- *Fourth Cambridge radio catalogue* by Gower *et al.* (1967) (7053) (4800 sources);
- *Molonglo reference catalogue of radiosources* by Large *et al.* (1981) (7029) (12 000 sources);
- *Catalogue of extragalactic radiosources having flux densities greater than 1 Jy at 5 GHz* by Kuehr *et al.* (1981) (7078) (3900 objects).

For X-ray observations there are the following catalogues on magnetic tape:

- *Third Uhuru X-ray catalogue* by Giacconi *et al.* (1974) (7012) (160 sources);
- *Fourth Uhuru X-ray catalogue* by Forman *et al.* (1978) (7018) (340 sources);
- *Second Ariel X-ray catalogue* by Cooke *et al.* (1978) (7019) (100 sources);
- *The Ariel (3A) catalogue of X-ray sources* by McHardy *et al.* (1981) (7058) (250 sources);
- *The HEAO A-1 X-ray source catalogue* by Wood *et al.* (1984) (7080) (840 sources).

7.6   Availability of catalogues

As a final note, it may be added that practically all the objects listed in this section – be they stars or galaxies – are included in the SIMBAD data base of the CDS Strasbourg. They are listed with their cross-identifications and main observational parameters. Bibliographic references to objects can also be found for stars from 1950 on and for non-stellar objects from 1982. This data base, and more specifically its bibliographic file, should always be consulted when one is interested in particular objects.

The foregoing summary represents the state of art at the time of writing – mid 1987. It is clear that the value of such a summary is strictly limited in time, and the reader is advised to consult the latest issues of the Data Centre publications (see Chapter 9) to get an update of this list. The field is constantly changing because old catalogues are replaced by new ones and new catalogues are added. At the present time, about five hundred catalogues exist and this number grows by about 10% per year.

7.7   Deficiencies of the present coverage

A perusal of the previous section shows the obvious fact that the coverage is very uneven. There are well-organized areas, such as stellar photometry and variable stars, and others where the coverage is sketchy or non-existent. The explanation of this lies in the organization of the data of each field. Traditionally – let us say fifty years ago – a catalogue was strictly the work of one person, most frequently that of a specialist who, near the retirement age, published his own files and, in this way, left a legacy for the next generation. This produced, in general, catalogues of high quality but, understandably, the author was unable to produce an updated edition of his catalogue. Data in this area were thus left without adequate coverage until a new specialist decided to redo a catalogue. Very often the principles under which the new edition was done differed from the ones for the previous edition, so that a certain heterogeneity inevitably resulted.[1]

Such practices are unsatisfactory when the data flow is large. Assume that a catalogue made at epoch T_1 contains N_1 objects and that there is an annual increase of n_1 papers carrying (on the average) data for m_1 stars. Leaving aside the fact that n_1 also increases with time (see Chapter 8), we have thus at epoch T_2 an increase of $n_1 m_1 (T_2 - T_1)$ over the N_1 stars of the first catalogue. Just to provide an example, we take the case of uvby photometry. As Table 5 from Chapter 8 shows, the annual increase in data is about 1000 for a total of 21 000, so that in five years the increase is about 25%. Since papers on photometry contain a rather small number of stars (see Mermilliod 1984, Table 1) namely about 60, we see that the user of the catalogue would have to consult about 80 papers to update the coverage. Obviously, very few astronomers can permit themselves the luxury to search the literature and locate these 80 papers and afterwards to scrutinize all of them, just to find out if the one star they are interested in has been observed. We find, therefore, the nice situation that data may exist but are not used because they are bibliographically unaccessible, i.e. buried in the literature. It is clear that at this point the original catalogue must be updated urgently.

If an astronomer wants to update his own catalogue, he is obliged to devote a significant part of his time to the new edition and, since data increase rapidly with time, the interval between successive updates tends to decrease. In this game the catalogue-maker finds himself in a difficult position, because he is 'running up-hill'. A second fact that tends to make things more difficult is that 'cataloguing' is regarded by many scientists as 'second class science'. Of course, it does not agree with the glamorous picture of the scientist working on research frontiers but, on the other hand, it would be foolish to stop cataloguing because we would be lost in a very short time in an ocean of published but otherwise unaccessible data.

Some of these difficulties may be solved by changing some of the assumptions that went into earlier work. One assumption was that the catalogue represented the final objective; in this way, the data collection (i.e. the data bank) is a means toward this goal. But one may certainly put things the other way around, so that the data bank is the main objective and the catalogue a by-product. The second assumption was that a catalogue was the by-product of research. One may also reverse this and say that research is based on data banks, so that their development is a prerequisite for further work. Both points become very important when it is not an isolated astronomer, but a team that is doing the work.

If this is the case, cataloguing can be regarded as part of the normal activities of the group, to be carried out in a systematic way. Each of the n members of the group can then devote a fraction p of his time to cataloguing or, to express it better, to 'working on the data bank'. The existence of such a situation is the explanation of the 'well-organized' areas such as variable stars and photometry. As an example, let us consider the Observatory of Lausanne group, under the leadership of Prof. B. Hauck. This group started working in 1973 and, over 13 years, has produced 21 photometric compilation catalogues, in which 15 scientists (including graduate students) have participated. Moreover, some of these catalogues, such as the one on Strömgren (uvbyβ) photometry, had regular updates, in 1973, 1975, 1980 and 1985. As to the percentage of time devoted to cataloguing, it is estimated at roughly 50% (Hauck, private communication). The work of this team has produced an excellent coverage of ground-based photometry (range 3000 Å–10 000 Å) and has contributed to an ever-increasing interest and use of photometric data.

Obviously, such a solution is needed most in areas where data growth is fast (see Chapter 8), whereas it is less urgent in areas of slow data growth.

7.8 Inadequately-covered areas

If we turn next to areas that are less well-covered, we find basically two different situations. First of all, it might be that a group of scientists is not interested in creating a data base. This is the case for radio-astronomical data. An enquiry by Westerhout and Jaschek (1977, 1978) has shown that 17 out of 23 radio observatories consulted were not interested in the centralization of information. Radio astronomers felt that each specialist knows where to get the material he needs. This is true, but is valid only for specialists, which constitute only a small fraction of all astronomers. An outsider to radio astronomy who wants to use radio data is therefore obliged to address himself to a number of colleagues before finding out who has what and how he can get the information he wants. Next, what should he do if two different sets of data exist? Should he use one rather than the other or combine both? How? All these questions could be solved much more easily if a centre for radio data existed, which organizes existing information in a coherent set of files, and from which any user can get the data he wants.

To dispel the suspicion that this may be an expression of the author's prejudice against radio astronomers, let me add that the same situation existed in optical astronomy not very long ago. So one may guess that it will be only a matter of time before a radio data centre is set up, and the earlier this is done the better it is for astronomy.[2]

The existing radio data have been surveyed recently in a meeting on 'Sky surveys' (*Inf. Bull. CDS No. 28*, 1985) and in a workshop on 'Surveys of the Southern galaxy' edited by Burton and Israel (1983).

A perplexing situation also exists in astrometry, for stellar proper motion and positions. This seems paradoxical because, since the 1840s a large part of the effort of the astronomical community has gone into the observation of star positions, which should in due time allow the determination of proper motions. Effectively, as we have seen, Sevarlic *et al.* (1978) lists more than 2×10^3 positional catalogues, and more than 4×10^2 exist in machine-readable form. But it is also clear that a non-specialist cannot use these observational data directly; the combination of the different catalogues has to be done by a specialist. Now, it turns out that only two catalogues – the SAO and the AGK3 – provide such averaged data for a large number of stars. Jaschek (1988) has shown that these are the two most popular catalogues in astrometry, a fact which underlines the necessity we all have for 'averaged data' catalogues in any field outside the one in which we are ourselves specialists.

In other areas with a poor catalogue coverage, a second type of situation exists, due to the complexity of data. For example, we have no catalogue of the chemical element abundances in stars, because of the large number of elements and the complexities (and uncertainties) of the theory needed to arrive at the abundances. The most we have is a bibliographic catalogue of 'metal abundances' (in stars, the word 'metal' being taken to imply all elements other than hydrogen and helium).

As another example we have interstellar clouds. Their forms and locations are so indeterminate that we have very few catalogues. There is, for instance, not even one on the strength of interstellar CaII absorption lines, despite the fact that these features were discovered 80 years ago. It is clear that only the work of a well-organized team may rescue this chapter of astronomy from its present state of underdevelopment.

Notes on Chapter 7

As remarked, the list of 'latest' catalogues is likely to be outdated very soon and must be completed by the reader, with the help of the publications indicated in the text.

1. The section on deficiencies in the coverage represents the author's personal views and might be perceived differently by others. The essential point is that the coverage is extremely uneven. As a rule, one may say that catalogues cover well

those branches of astronomy for which well-established techniques of observation and reduction are available. Fields in which these two conditions are not fulfilled are worse off, as was the case in stellar photometry before 1960. At about this time, observational and reduction techniques became standardized and catalogues followed almost immediately. If astronomers do not agree on a technique and on reduction procedures the area remains in a state of flux until the community reaches a consensus. In such cases, the information that is vital is a detailed bibliography for each object (see Chapter 10 for more details). Such information may to some degree replace the missing catalogue(s).

2. For a possible improvement see the papers in *BICDS* **35**.

References

Aaronson, M., Huchra, J., Mould, J. R., Tully, R. B., Fischer, J. R., Worden, H. van, Mebold, U., Siegman, B., Berriman, G. and Persson, S. E. (1982) *Ap. J. Suppl.* **50**, 24

Abt, H. A. and Biggs, E. S. (1972) *Bibliography of stellar radial velocities*, Latham, New York

Acker, A., Marcout, J. and Ochsenbein, F. (1981) *AA Suppl.* **43**, 265

Ardeberg, A. and Virdefors, B. (1980) *AA Suppl.* **40**, 3071

Argelander, R. (1859–62) *Bonner Sternwarte Beobachtungen*, 3, 4 and 5

Arp, H. C. (1966) *Ap. J. Suppl.* **14**, 1

Baiesi-Pillastrini, G. C., Palumbo, G. G. C. and Vettolani, G. (1984) *AA Suppl.* **56**, 363

Barbier, M. (1975) CDS catalogue 3038

Barbier-Brossat, M. and Petit, M. (1986) *AA Suppl.* **65**, 59 (= CDS catalogue 3118)

Barbieri, C., Capaccioli, M., Custiani, S., Nardo, G. and Omizzolo, A. (1983) *Mem. Soc. Astron. Ital.* **54**, 601

Batten, A. H., Fletcher, J. M. and Mann, P. J. (1978) *Publ. Dominion Astrophys. Obs.* **15**, 121

Bennett, A. S. (1961) *Mem. RAS* **68**, 163

Bok, B. J. (1937) *The distribution of the stars in space*, University of Chicago Press

Breger, M. (1976) *Ap. J. Suppl.* **32**, 7

Breger, M. and Hsu, J. C. (1981) *BICDS* **23**, 51

Burbidge, G. R. and Crowne, A. H. (1979) *Ap. J. Suppl.* **40**, 3

Burton, W. B. and Israel, F. P. (1983) (ed.) *Survey of the southern sky*, Reidel

Cannon, A. J. and Pickering, S. (1924) *Harvard Annals* **91–100**

Cayrel de Strobel, G., Bentolila, C., Hauk, B. and Duquenoy, A. (1984) CDS catalogue 3096

Clark, D. H. and Caswell, J. L. (1976) *MN* **174**, 267

Collins, M. (1977) *Astronomical catalogues 1951–1975*, London INSPEC Bibl., Ser. 2

Cooke, B. A., Ricketts, M. J., Maccacaro, T., Pye, J. P., Elvis, M., Watson, M. G., Griffiths, R. E., Pounds, K. A., McHardy, I., Maccagni, D., Seward, F. D., Page, C. G. and Turner, M. J. L. (1978) *MN* **182**, 489

Cucchiaro, A., Jaschek, M. and Jaschek, C. (1979) *BICDS* **17**, 93

Davis, R. J., Deutschman, W. A., and Haramundanis, K. L. (1973) *Smithsonian Astrophysical Observatory Special Reports*, No. 350

Didelon, P. (1983) *AA* **53**, 119

Dixon, R. S. (1976) In *IAU Coll. 35, Compilation, critical evaluation and distribution of stellar data*, Jaschek, C. and Wilkins, G. A. (ed.), Reidel, p. 167

Duus, A. and Newell, B. (1977) *Ap. J. Suppl.* **35**, 209

Eichendorff, W. and Reinhardt, M. (1979) *ASpS* **61**, 153

Eichhorn, H. (1974) *Astronomy of star positions*, Ungar, New York

Evans, D. S. (1967) In *IAU Symp. 30, Determination of radial velocities and their applications*, Batten, A. H. and Heard, J. F. (ed.), Academic Press

Fairall, A. P., Lowe, L. and Dobbie, P. J. K. (1985) *Publ. Dep. Astr. Cape Town*, No. 5

Fernandez, A., Mathez, G. and Nottale, L. (1984) *Comptes Rendues Journées de Strasbourg*, **VI**, p. 33, Observatoire de Strasbourg

Forman, W., Jones, C., Cominsky, L., Julien, P., Murray, S., Peters, G., Tananbaum, H. and Giacconi, R. (1978) *Ap. J. Suppl.* **38**, 357

Fracassini, M., Pasinetti, L. E. and Manzolini, F. (1981) *AA Suppl.* **44**, 155

Fresneau, A. (1978) *BICDS* **15**, 67

Fresneau, A. (1981) *BICDS* **20**, 110

Gezari, D., Schmitz, M., Mead, J. M. (1982) *NASA Technical Monographs 83819*

Giacconi, R., Murray, S., Gursky, H., Kellogg, E., Scheier, E., Matilsky, T., Koch, D. and Tananbaum, H. (1974) *Ap. J. Suppl.* **27**, 37

Giclas, H. L., Burnham, R. and Thomas, N. G. (1978a) *Lowell Obs. Bull.* **89** (1958) to **136** (1966) (= CDS catalogue 1079)

Giclas, H. L., Burnham, R. and Thomas, N. G. (1978b) *Lowell Obs. Bull.* **164**

Gill, D. and Kapteyn, J. C. (1895–1900) *Cape Photographic Durchmusterung, Cape Ann.* **3** (1896), **4** (1897), **5** (1900)

Gliese, W. (1983) In *IAU Coll. 76, Nearby stars and the stellar luminosity function*, Davis Philip, A. G. and Upgren, A. (ed.), Davis Press, New York

Gliese, W. and Jahreiss, H. (1979) *AA Suppl.* **38**, 423

Golay, M. (1974) *Introduction to astronomical photometry*, Reidel

Gower, J. F. R., Scott, P. F. and Wills, D. (1967) *Mem. RAS* **71**, 49

Gunn, J. E. and Stryker, L. L. (1983) *Ap. J. Suppl.* **56**, 278

Haramundanis, K. L. (1966) *Smithsonian Astrophysical Observatory star catalog*, Smithsonian Institution, Washington DC

Hauck, B. and Mermilliod, M. (1985) *AA Suppl.* **60**, 61

Heck, A., Egret, D., Jaschek, M. and Jaschek, C. (1984) *IUE Low-dispersion spectra reference atlas, Part I: Normal stars*, ESA SP-1052

Heckmann, O. and Dieckvoss, W. (1975) *AGK3*, Hamburg

Hoffleit, D. and Warren, W. (1986) *Bright star catalogue, fifth edition* (= CDS catalogue 5050)

Houk, N. (1978) *Michigan Obs. Publ.* **II**

Houk, N. (1982) *Michigan Obs. Publ.* **III**

Houk, N. and Cowley, A. P. (1975) *Michigan Obs. Publ.* I

Humphreys, R. M., McElroy, D. B. and Ahigo, K. (1984) *Catalogue of stars in stellar associations and young clusters*, University of Minnesota

Huth, H. and Wenzel, W. (1986) 'Bibliographic catalogue of variable stars', *BICDS* **4**, 11 and *BICDS* **20**, 105

Huth, H. and Wenzel, W. (1986) CDS catalogue 6038

Joint IRAS Science Working Group (1984) Jet Propulsion Laboratory Explanatory Supplement

Jacoby, G. H., Hunter, D. A. and Christian, C. A. (1984) *Ap. J. Suppl.* **56**, 278

Jamar, C., Macau-Hercot, D., Monfils, A., Thompson, G. I., Houziaux, L. and Wilson, R. (1976) *ESA Special Report* **27**

Jaschek, C. (1984a) In *IAU Symp. 111, Calibration of fundamental stellar quantities*, Hayes, D. S., Pasinetti, L. and Davis Philip, A. G. (ed.), Reidel, p. 331

Jaschek, C. (1984b) In *The MK process and stellar classification*, Garrison, R. F. (ed.), David Dunlap Obs., Toronto, p. 94

Jaschek, C. (1988) In *IAU Symp. 133 Mapping the sky: past heritage and future directions*, Debartet *et al.* (ed.), Reidel

Jaschek, C., Conde, H., Sierra, A. C. de (1964) *Publ. La Plata Ser. Astr.* **28**, Part 2

Jaschek, M. (1978) *BICDS* **15**, 121

Jaschek, M. and Egret, D. (1984) *Catalogue of stellar groups, Part I*, Strasbourg

Jasniewicz, G. (1982) *AA Suppl.* **49**, 99

Jenkins, L. F. (1963) *General catalog of trigonometric stellar parallaxes*, and *Supplements*, Yale University Observatory

Kalinkov, M., Stavrev, K. and Kaneva, I. (1979) Tape prepared by the Bulgarian Academy of Science (= CDS catalogue 7004)

Karachentsev, I. D. (1986) CDS catalogue 7083

Karachentsev, I. D., Lebedev, V. and Shchervanovskij, A. (1972) *BICDS* **29**, 87

Kennedy, P. M. (1983) *MK classification catalogue*, Mt. Stromlo and Siding Spring Observ., Australia

Kobold, H. (1926) In *Enzyklopaedie der mathematischen Wissenschaften*, Vol. VI, 28, Teubner Publ. Co., Leipzig, p. 241

Kogoshvili, N. G. (1983) *BICDS* **25**, 63

Köppen, J. (1988) To be published

Kraicheva, Z., Popova, E., Tutukov, A., Yungelson, L. (1980) *BICDS* **19**, 71

Kukarkin, B. V., Kholopov, P. N., Efremov, Yu. N., Kukarkina, N. P., Kurochkin, N. E., Medvedeva, G. I., Perova, N. B., Fedorovich, V. P., Frolov, M. S. (1971) *General catalogue of variable stars*, Moscow State University

Kukarkin, B. V. *et al.* (1981) *Supplement to the general catalogue of variable stars*, Moscow State University

Kukarkin, B. V. *et al.* (1985) *General catalogue of variable stars, fourth edition, updated to Orion*, Moscow State University

Kuehr, H., Witzel, A., Pauliny-Toth, I. I. K. and Nauber, U. (1981) *AA Suppl.* **45**, 367

Lanz, T. (1986) *AA Suppl.* **65**, 195

Large, M. I., Mills, B. Y., Little, A. G., Crawford, D. F. and Sutton, J. M. (1981) *MN* **194**, 693

Luyten, W. J. (1979, 1980) *NLTT Catalog, Vols. I–VI*, University of Minnesota

Lynds, B. (1962) *Ap. J. Suppl.* **7**, 1 and updates

Lynds, B. (1965) *Ap. J. Suppl.* **12**, 163

Lynga, G. (1987) *Catalogue of open clusters, fifth edition* (= CDS catalogue 7092)

Macau-Hercot, D., Jamar, C., Monfils, A., Thompson, G. I., Houziaux, L. and Wilson, R. (1978) ESA Special Report **28**

McHardy, I. M., Lawrence, A., Pye, J. P. and Pounds, K. A. (1981) *MN* **197**, 893

Magnenat, P. (1976) *BICDS* **11**, 17

Markarian galaxies (1982) Collection of lists published by Markarian *et al.* from 1967–81 (= CDS catalogue 7061)

Marshalkova-Polechova, P. (1974) *ASpS* **27**, 3

Mathewson, D. S., Ford, V. I., Klare, G., Neckel, Th. and Krautter, J. (1978) *BICDS* **14**, 115

McClure, R. D. and Forrester, W. T. (1981) *Publ. Dominion Astrophys. Obs.* **15**, 439

Mennessier, M. O., Gomez, A., Creze, M. and Morin, D. (1978) *BICDS* **15**, 83

Mermilliod, J. C. and Nicolet, B. (1977) *AA Suppl.* **29**, 259

Mermilliod, J. C. (1979) *BICDS* **16**, 2

Mermilliod, J. C. (1983) *BICDS* **25**, 79

Mermilliod, J. C. (1984) *BICDS* **26**, 3

Mermilliod, J. C. (1985) Update of *AA Suppl.* **26**, 419 (1976) (= CDS catalogue 4017)

Mermilliod, J. C. (1987) *AA Suppl.* **71**, 413

Mitchell, K. J., Brotzman, L. E., Warnock, A. and Usher, P. D. (1982) *ASpS* **88**, 219

Morel, M. and Magnenat, P. (1978) *AA Suppl.* **34**, 477

Nicolet, B. (1975) *AA Suppl.* **22**, 239

Nicolet, B. (1978) *AA Suppl.* **34**, 1

North, P. (1984) *BICDS* **27**, 133

Ochsenbein, F. (1974) *AA Suppl.* **15**, 215

Ochsenbein, F., Bischoff, M. and Egret, D. (1981) *AA Suppl.* **43**, 259

Ochsenbein, F. (1984) *BICDS* **26**, 75

Pedoussaut, A., Capdeville, A., Ginestet, N. and Carquillat, J. M. (1986) CDS catalogue 4016

Popper, D. M. (1980) *Ann. Rev. Astron. Astrophys.* **18**, 115

Ristenpart, F. (1901) In *Handwörterbuch der Astronomie*, Valentiner, W. (ed.), Trewendt Co., Breslau, Vol. 3 second part, *Sterncataloge und Karten*

Rood, H. J. (1980) CDS catalogue 7036

Rufener, F. (1981) *AA Suppl.* **45**, 207

Ruprecht, J., Balazs, B. and White, R. E. (1983) *Soviet Astronomy* **27**, 358

Sandage, A. and Tamann, G. A. (1981) *A revised Shapley-Ames catalog of bright galaxies*, Carnegie Inst. Washington, Publ. 635

Schönfeld, E. (1886) *Bonner Sternwarte Beobachtungen* **8**

Seal, R. A. and Martin, S. S. (1982) A bibliography of astronomy 1970–79. Libraries Unlimited Inc., Littleton, Colorado

Seiradakis, J. H. (1986) CDS catalogue 7008

Sevarlic, B. M., Teleki, G. and Szdeckzy-Kardoss, G. (1978) *Epitome fundamentorum astronomiae. Pars I, Catalogues of star positions*. Publ. Dep. Astr. Belgrade No. 7

Sevarlic, B., Teleki, G. and Knezevic, Z. (1982) *Epitome fundamentorum astronomiae, Pars II*, Publ. Obs. Belgrade **29**, 71

Sharpless, S. (1959) *Ap. J. Suppl.* **4**, 257

Sulentic, J. W. and Tifft, W. G. (1974) *The revised new general catalog of nonstellar astronomical objects, First edition*. University of Arizona Press

Svechnikov, M. A. and Bessonova, L. A. (1984) *BICDS* **26**, 99

Thome, J. M. (1892–1932) Córdoba Durchmusterung, Resultados Obs. Córdoba **16**, **17**, **18** and **32**

Thompson, G. I., Nandy, K., Jamar, C., Monfils, A., Houziaux, L., Carnochan, A. and Wilson, R. (1978) *Catalogue of stellar ultraviolet fluxes*, Science Research Council, London

Uesugi, A. and Fukuda, I. (1982) *Revised catalog of stellar rotational velocities*, Kyoto University, Japan

van den Bergh, S. (1966) *AJ* **71**, 990

Vaucouleurs, G. de and Vaucouleurs, A. de (1964) *Reference catalog of bright galaxies, First edition*. University of Texas Press

Vaucouleurs, G. de and Vaucouleurs, A. de (1976) *Reference catalog of bright galaxies, Second edition*. University of Texas Press

Veron-Cetty, M. P. and Veron, P. (1987) New version of ESO Sci. Rep. No. 1 and No. 4

von der Pahlen, E. (1937) *Lehrbuch der Stellarstatistik*. Barth, Leipzig

Vorontsov-Velyaminov, B. A. and Arkhipova, V. (1962–74) *Sternberg Inst.* Vol. I–IV, Moscow

Weedman, K. W. (1977) *Ann. Rev. Astron. Astrophys.* **15**, 69

Weedman, K. W. (1978) *MN* **184**, 11

Wendker, H. J. (1984) *Abh. Hamburger Sternw.* **10**, 1 and update to 1984

Wesselius, P. R. (1979) NASA Goddard Status Report on machine-readable catalogs, Oct. 1986 edition

Wesselius, P. R., Duinen, R. J., de Jonge, A. R. W., Aalders, J. W. G., Luinge, W. and Wildeman, K. J. (1982) *AA Suppl.* **49**, 427

Westerhout, G. and Jaschek, C. (1977, 1978) *BICDS* **13**, 28 (1977), **14**, 20 and **15**, 99 (1978)

Wilson, R. F. (1953) *General catalog of stellar radial velocities*, Carnegie Institute, Washington, Publ. 601

Wood, F. B., Oliver, J. P., Florkowski, D. R. and Kich, R. H. (1980) *Publ. Dept. Astronomy*, Florida, Vol. 1

Wood, K. S., Meekins, J. F., Yentis, D. J., Smathers, H. W., McNutt, D. P.,

Bleach, R. D., Byram, E. T., Chubb, T. A. and Friedman, H. (1984) *Ap. J. Suppl.* **56**, 507

Worley, C. E. and Heintz, W. D. (1983) *Publ. Naval Obs. Washington 2nd ser.* **24**, 7

Worley, C. E. and Douglass, G. E. (1984) *BICDS* **28**, 165

Zwicky, K. *et al.* (1961, 1963, 1965, 1966, 1968) *Catalog of galaxies and of clusters of galaxies, Volumes I–VI*, California Institute of Technology, Pasadena

8

The growth of data

The amount of data increases with time, as everybody knows. This is part of the general growth in information and documentation, which we shall examine in more detail in Chapter 10. Here we shall concentrate on the quantity of (analysed) data, as a function of time. This gives some idea of how much data should be available to the modern scientist, and thus defines the size of data banks.

8.1 Growth characteristics

We start by introducing some parameters to characterize the growth. Call $n(t_i)$ the number of data of a certain type available at the epoch t_i (in years). We can define the growth rate.

$$\Delta n = n(t_2) - n(t_1); \qquad \Delta t = t_2 - t_1;$$

$$\alpha = \frac{\Delta n}{\Delta t},$$

α being, in general, a function of time, with the obvious condition $\alpha > 0$. If α grows (diminishes) with time, we shall speak of increasing (decreasing) growth. If α is constant, the number of data increases linearly; if α grows exponentially, $n(t)$ also grows exponentially. In such a case we may define

$$\beta = \frac{\log n(t_2) - \log n(t_1)}{t_2 - t_1}.$$

To derive either α or β, a certain number of values of $n(t_i)$ must be known; if the data are few, obviously the variations in α can only be established approximately.

Another way to characterize exponential increase is by indicating the number of years needed to double the number of data. If this period is called P(years) one has $P = 0.301/\beta$.

Usually, objects are measured independently more than once in a given period. If $H(t_i)$ is the total number of data of a certain kind published up to epoch t_i, we can define an 'index of redundancy'

$$f = \frac{H(t_i)}{n(t_i)}.$$

A large value of f implies that many objects are measured many times. For some objects this is done because of suspected variability or because they are chosen as standards but, more often, it happens because observers are unaware of previous measurements, i.e. because of bibliographic inaccessibility.

8.2   The growth of astronomical information

The increase of data with time manifests itself in different ways. So, one may consider:

(a) one given datum, studied for different objects, such as the number of galaxies with measured redshift;
(b) the number of items of information of different kinds available for a given object (parameters characterizing a star, for instance);
(c) the number of different types of astronomical objects available for study.

The statistics that are easiest to perform are those of type a because the situation is very clear: an object has been or has not been measured. Statistics of type b and c are more difficult, because the situations are complex and so are the answers. We shall thus start with statistics of type a, and comment later on types b and c.

8.2.1   *One datum for different objects*

We shall examine the growth of certain kinds of astronomical data, chosen mostly from fields where the number of objects that can be studied is almost unlimited, to avoid complications due to a finite sample.

The values of $n(t_i)$ used come mostly from printed catalogues, with the publication year taken as t_i. Although this is usually one or two years later than the epoch at which the bibliographic survey for the catalogue was completed, in many cases we have no information about the 'closing date' of the bibliographic survey. For this reason we have used the publication year.

In what follows, numbers in parentheses are uncertain, either because

Table 8.1. *Number of asteroids having known orbits*

t_i	n_i	α	β	t_i	n_i	α	β
1859	42			1929	1116	22	
1869	90	5		1939	1489	37	
1879	169	8		1949	1565	8	
1889	267	10	0.022	1959	1626	6	0.0089
1899	444	18		1969	1735	11	
1909	569	12		1979	2188	45	
1919	895	33		1985	3259	108	

Note: n_i =number of asteroids, t_i = epoch, α and β – see text.
Sources: Connaissance des Temps; From 1919 on, *Astronomischer Jahresbericht*;
from 1959 on, *Ephemerides of minor planets*; 1979 and 1985 *Minor planet circulars*
Nos. 5036 and 9689.

the author of the catalogue did not quote figures (when they were
estimated by this author) or because different data are given in the sources,
which have had to be unscrambled.

Many numbers collected in the following tables are open to the criticism
that catalogues do not usually provide consistent values of $n(t_i)$. This has
various causes: either the author discarded some previously published
data, or he did not achieve or attempt completeness, or in the different
catalogues on a given subject the different authors did not include exactly
the same kind of data. Therefore, all figures should be considered to be
approximate, with a tentative uncertainty of the order of $\pm 10\%$. The α and
β values are thus less accurate than one could wish.

Asteroids. We start with an example from solar-system astronomy, which
falls outside the field of this book but is very illustrative because it covers
180 years (Table 8.1). The number of asteroids known at each epoch was
taken to be equal to the number of objects having a well-defined orbit.
Although requirements for the latter became stricter in 1935, I have felt
that it is difficult to take this factor into consideration, and I have thus
ignored this change.

The first asteroid was discovered in 1801. The growth was exponential
until 1920, with a 5% yearly increase. From 1949 on (i.e. after the Second
World War) it grew again exponentially with a 2.2% annual increase.

Stars having spectroscopic binary orbits. The data are given in Table 8.2.

Table 8.2. *Number of known spectroscopic binary orbits*

	t_i	n_i	α
Campbell	1910	70	
Moore	1924	248	13
Moore and Neubauer	1945	480	11
Batten	1967	737	12
Halbwachs (private comm.)	1981	1121	27
Carquillat (private comm.)	1986	1300	36

Note: t_i = epoch, n_i = number of stars having at least one orbit, α – see text.

Table 8.3. *Trigonometric parallaxes*

	t_i	n_i	α	f
Started around	1842			
Schnauder	1923	(1200)		
Schlesinger & Jenkins	1935	(4144)	240	
Jenkins	1952	5822	97	
Jenkins	1963	6399	52	1.0

Note: t_i = epoch, n_i = number of stars with measured parallax, α and f – see text.

Table 8.4. *Visual binaries*

	t_i	n_i	α
Started around	1820		
Innes	1927	2.7×10^4	
Aitken	1932		
Jeffers, van den Bos & Greeby	1963	6.42×10^4	1300
Worley	1976	7.0×10^4	460
Worley and Douglass	1984	7.4×10^4	500
Orbits			
van den Bos	1926	118	
Baize	1950	253	6
Worley	1963	536	22
Finsen & Worley	1970	696	23
Worley and Heintz	1983	928	18

Note: Only good quality orbits were included. t_i = epoch, n_i = number of known double stars (or orbits), α – see text.

After constant growth until the sixties, the growth is accelerating rapidly.

Stars having trigonometric parallaxes determined. The data are gathered in Table 8.3.

The growth is decreasing with time.

Visual binaries. The data are collected in Table 8.4.

The growth is decreasing for the number of visual binaries and about constant for the orbits known.

Photometric data. Stars having photoelectric measurements in different systems are listed in Table 8.5.

Table 8.5. *Stars measured photometrically in different systems*

	t_i	n_i	α	f
UBV				
Started by				
Johnson & Morgan	1953	2.9×10^2		
Blanco *et al.*	1968	2.4×10^4	1600	1.4
Mermilliod	1977	5.3×10^4	4100	1.4
Mermilliod	1984	7.2×10^4	2700	
			$\beta = 0.036$	
uvby				
see Hauck (1985)	1965	1.2×10^3		
	1973	7.5×10^3	900	
	1975	9.3×10^3	900	1.4
	1977	1.44×10^4	2500	
	1978	1.9×10^4	4600	
	1984	4.3×10^4	4000	
			$\beta = 0.073$	
Geneva				
see Hauck (1985)	1964	3.4×10^2		
	1966	6.8×10^2	170	
	1971	1.4×10^3	140	
	1976	4.6×10^3	640	
	1978	1.3×10^4	4200	
	1980	1.5×10^4	1000	
	1982	2.2×10^4	3500	
			$\beta = 0.097$	

Note: t_i = epoch, n_i = number of stars measured in the system; α, β and f – see text.

The growth seems constant or decelerating in UBV and uvbyβ, whereas it is fully exponential in Geneva photometry.

Variable stars. The number of known variable stars is given in Table 8.6.

The growth was exponential with an average increase of about 7% annually, up to 1933. Afterwards, the increase fell to 3.2%. The first seven data were taken from Hoffmeister (1970).

Pulsars. We have chosen pulsars as representative of objects discovered very recently. The growth history is given in Table 8.7.

The increase is exponential, with an annual growth of 20%! Such rapid growth is however unlikely to continue; in 1987 the total number is not 10^3 as expected but more like 500.

Quasars. We have examined as a second example of recently discovered objects the quasars, i.e. quasi-stellar radio sources. They were discovered in the nineteen-sixties; probably one can regard Schmidt's (1963) paper as the one identifying unambiguously the first quasar. Afterwards the number of known quasars grew rapidly as can be seen in Table 8.8, although a certain deceleration seems indicated after 1983.

We may now attempt to summarize the results from Tables 8.1 to 8.8. We find the growth to be decreasing (trigonometric parallaxes), constant (visual binary orbits), growing (spectroscopic binary orbits) and exponential (asteroids, photometric data, variable stars, pulsars and quasars). The statement that data growth is exponential, voiced by some authors, is thus not true in general. As for the individual values of α and β, they vary widely. For β in particular one finds values ranging from 0.0089 (asteroids) to 0.08 (quasars, pulsars) and 0.113 (Geneva photometry) implying doubling periods between 33 and 3 years. One may conclude that data in the areas of general interest of today's astronomers, are growing quickly – thus, in general, data growth reflects the interests of the scientific community. We may see that in more detail with the following crude model. Assume that the output of a certain kind of data per year per scientist is δ, and let us assume that, on a short timescale, δ is constant. Call m the number of active astronomers in this field. The annual growth rate of data is then

$$A = \delta \cdot m.$$

Of these two quantities, δ almost never diminishes (unless, for instance,

Table 8.6. *Variable stars*

	t_i	n_i	$\log n_i$	β
Argelander	1850	24	1.38	
Schönfeld	1875	143	2.16	
Chandler	1896	393	2.59	0.028
Müller & Hartwig	1915	1 687	3.23	
Heise	1923	2 233	3.35	
Prager	1933	5 826	3.76	
Schneller	1943	9 476	3.98	
Kukarkin & Parenago	1948	10 912	4.04	0.011
Kukarkin *et al.*	1958	14 711	4.17	
Kukarkin *et al.*	1970	20 437	4.31	
Kholopov *et al.*	1985	28 450	4.45	

Note: t_i = epoch, n_i = number of known variable stars, β – see text.

Table 8.7. *Pulsars discovered*

	t_i	n_i	$\log n_i$	β
First discovery	1967	1	0	
Hewish	1970	44	1.64	0.08
Manchester and Taylor	1977	149	2.17	
Manchester and Taylor	1981	330	2.52	

Note: t_i = epoch, n_i = number of pulsars known, β – see text.

Table 8.8. *Number of quasars with measured redshifts*

	t_i	n_i	β
Schmidt	1963	1	
Burbidge	1967	101	
de Veny *et al.*	1971	202	0.0806
Burbidge *et al.*	1977	633	
Hewitt and Burbidge	1980	1491	
Barbieri *et al.*	1983	2004	
Veron-Cetty and Veron	1984	2251	
Veron-Cetty and Veron	1985	2835	

Note: t_i = epoch, n_i = number of quasars, β – see text.

the instrument stops or is destroyed, which may happen in the case of observations from space). It can increase rapidly if a new technique is introduced. This is nicely illustrated in the case of spectroscopic binary orbits, where the sudden acceleration in the nineteen-seventies (see Table 8.2) is related to the use of a new type of apparatus for the measurement of stellar radial velocities (the radial spectrometer), with which accurate radial velocities can be measured quickly.

The quantity m, on the other hand, is highly variable: when a new type of object is discovered, there is a sudden rush of outsiders to this particular field. As a consequence the product A increases rapidly. The case of exponential growth can be explained by the assumption that, in some fields (for instance in variable star research), the number of active astronomers (m) is a constant proportion of the astronomical community, which itself grows exponentially. This presupposes that the community as a whole considers such a field to be sufficiently interesting. This suggests, on the other hand, that decreasing or constant growth rates demonstrate a diminishing interest of the community in certain fields. Such an attitude often reflects the lack of a technological innovation, as in the field of radial velocities before the nineteen-seventies.

One may, however, doubt whether exponential growth in science lasts forever. De Solla Price (1963, 1975) has shown that all exponential growth in science is doomed since there is a ceiling in science budgets and manpower. This implies that exponential growth is but one stage in growth and that one must look for more general expressions of growth. Growth functions have been investigated by biologists who have shown that a very general form is the logistic or sigmoid curve given by

$$f(t) = (1 + \exp(a(t - t_0)))^{-1}$$

which represents the number of individuals as a function of time. Fig. 8.1 shows a graph of the function. The curve is composed of three parts: a slow linear increase at first, a rapid exponential increase, and a third part that represents saturation. This part shows a diminishing growth, tending asymptotically to a limiting value. The exponential growth region can be delimited by the values $a(t - t_0) = \pm 2$; these values correspond to the intersection of the osculating tangent with $f = 0$ and $f = 1$. At both extremes, the difference between the straight line and the curve is only 12%. In Fig. 8.2 the same growth function is represented on a logarithmic graph. We find again an exponential region, preceded and followed by

Fig. 8.1. The sigmoid curve (see text). In abscissae, a linear function of time.

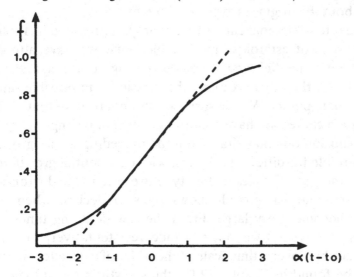

Fig. 8.2. The logarithmic sigmoid curve. In abscissae, a linear function of time.

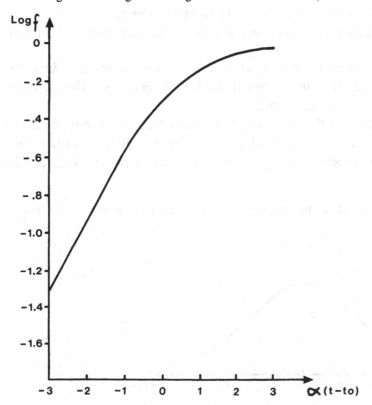

sections of slower growth. Finally in Fig. 8.3 the growth rate is given, which is symmetrical about the origin, as expected from Fig. 8.1.[1]

One can try next to fit sigmoid curves to the samples provided in Tables 8.1–8.8. In the cases of asteroids and variable stars, we have already indicated that the exponential growth was decreasing, a fact that adapts easily to Fig. 8.2. For the other six cases, the decision is impossible since there are too few data points. We can surmise – but not prove – that in the long run all the data curves will have a saturation level or ceiling.

The main conclusion one may draw from the preceding is that growth rates are very variable for different subjects, with exponential growth for subjects of high interest. (Topics of this type are often called 'frontier research'.) The exponential growth shows signs of decline when the number of data becomes very large. From the few doubling times we derived (4 years to 33 years) we may conclude very tentatively that the number of data doubles over a time scale of the order of a decade. Similar time scales will be found in Chapter 12 for the growth in the volume of publications.

8.2.2   *Information available for a given class of objects*

We shall start with a relatively simple example, namely information on stars.

The information that can be obtained for stars is summarized in Table 8.9. I have grouped the information under broad headings, although often the subdivisions are not clear-cut.

I present in Table 8.9, a summary of the information we may obtain on stars nowadays. I have grouped it according to physical rather than observational characteristics, just to make the meaning clearer. So

Fig. 8.3. Growth rate in the sigmoid curve. In abscissae, a linear function of time.

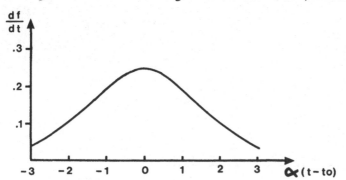

Table 8.9. *Information available on stars*

1. *Global, independent of star*
Position, proper motion, magnitude, radial velocity, distance

2. *Global, characterizing the star*
Mass, radius, luminosity, form, magnetic field, mass loss, age, internal mass
distribution, variability, binary system membership (orbit characteristics),
rotation

3. *Outer layers*
Photosphere (temperature, electron pressure, turbulence, colour (at different
wavelengths) composition, magnetic field, spots), chromosphere, corona (mass
loss, stellar winds), circumstellar matter

luminosity is included, but not absolute magnitude, and temperature
rather than spectral type or some multiband photometry. A difficulty that
comes up is that some information can be encoded in just one number (e.g.
mass) whereas others, such as 'corona' require the specification of quite a
number of parameters: dimensions, temperature (and its variations with
height), pressure, velocity fields etc. In conclusion, the information on just
one star may be very large, depending on the distance of the star and the
interest astronomers have in it; for instance, α Lyr is much more 'interesting' than α Cen, despite its larger distance. Obviously, not all types of
information listed are known for a given star – for most of them the
information is very scanty.

It is difficult to provide a parameter that measures the quantity of
information available on a given star. Taking Table 8.9 at face value one
could foresee some thirty types of information, and the number growing
over the years. At the start of this century, probably no more than ten types
were envisaged. Since such an increase is slow, the statistics of type *b* are
not very helpful for characterizing the growth of data; what they do is to
illustrate the progress of our knowledge.

If instead of stars one considers groups of stars, e.g. clusters, the
foregoing considerations apply equally. However, the number of types of
information increases considerably because, as well as describing each
cluster component (stars, dust, gas), we need to characterize the distribution of each component in six-dimensional phase space. Obviously, if
the number of components becomes large, either we may not be interested
in the detailed information on each object, or we may be forced to resign

from any attempt to obtain it, so that for complex systems like galaxies the number of information types grows much more slowly than the number of objects.

8.2.3   *Number of different kinds of astronomical objects*

We may study similarly the statistics of type c, i.e. the growth of the number of different types of astronomical objects. This was done by Harwit (1981) in his book *Cosmic discovery*. He lists 43 major astronomical discoveries of which one third are since 1955. These fifteen discoveries are infrared stars, cosmic masers, pulsars, infrared radio sources, unidentified radio sources, quasars, superluminous sources, X-ray stars and galaxies and clusters of galaxies, infrared galaxies, γ-ray showers and microwave, X- and γ-radiation background.

We may remark again that such listings of discoveries characterize the growth of astronomy rather than the growth of data, although it is true that they contribute to it. Clearly, each new type of object will be studied to obtain various types of information, like that listed in Table 8.9.

8.3   **Incompleteness of our knowledge of the Universe**

Summing up, we see that we have a three-dimensional growth of data: in the types of objects, in the number of parameters of each object and in the number of objects. A 'compound' growth-rate indicator characterizing the three-dimensional growth seems not to be available at the present time.

With an exponential growth of the number of subjects and of data one may address the question of how complete our knowledge of the surrounding Universe is. The figures given above may induce us to believe that we know essentially everything about the objects in the Universe, or at least in the nearby Universe.[2]

For the inventory of the solar system one may be optimistic, because exploration is proceeding rapidly – before our very eyes. But as a sobering fact, let us recall that, for instance, rings around planets – other than those of Saturn – were discovered only in the last decade. If such a thing can happen in our solar neighbourhood, one can only surmise that things will be worse farther out.

Let us start examining the sample of nearest stars, defined as those with distances up to 25 pc ($\approx 8 \times 10^{19}$ cm). Jahreiss (1985) indicates 2372 stars with measured trigonometric parallaxes and 421 for which only photometric and/or spectroscopic distance estimates exist. Among the 2800

Table 8.10. *Completeness of data for stars with M ≥ 6.5*

	1962	1982
Photoelectric magnitudes	50%	95%
B – V colours	50%	94%
MK spectral classes	75%	93%
Parallaxes (trigonometric)	30%	31%
Radial velocities	75%	93%
Rotational velocities	*	43%

* Not given.

stars, about three quarters have B, V photoelectric photometry. For stars brighter than V = 11 (which corresponds to M_v = 9) there are also spectral types, radial velocities and precise medium-band photometry. Beyond this limit, down to V = 19, the completeness of the data peters out rapidly.

Attention should also be paid to the fact that, according to Wielen *et al.* (1983), there should be 3875 objects within 25 pc; according to the figure quoted above we have recognized only two thirds of this number.

The same happens if one considers the sample of stars brighter than 6.5 mag, as given in the catalogue of 'Bright stars', third edition (Hoffleit 1962) and fourth edition (Hoffleit & Jaschek 1982). A comparison of the data in the two editions (Table 8.10) is rather instructive, because it shows convincingly how incomplete our knowledge of the basic parameters of even bright stars was two decades ago.

In 1982 the situation had improved – except for parallaxes and rotation. Of these, parallaxes will probably never be completed, because most stars are too far away to yield significant values. Rotational velocities, on the other hand, have to wait for better techniques; up to now we know rotational velocities only exceptionally for stars later than F-type. It should also be said that many of the existing data are of rather low quality, which should certainly be improved. And on top of this one may add that the situation is far from satisfactory in other respects. For instance, variability is known for 22% of stars, whereas the true percentage is probably twice as large. Chemical composition analyses have been done for about 10% of stars; ultraviolet data exist for 50%. The *Bright star catalog* quotes only

Table 8.11. *Data available on bright
galaxies, from de Vaucouleurs* et al. *(1976)*

Total number of galaxies: 4364	
Revised morphological types	87%
Yerkes types	15%
Hubble types	13%
Isophotal: diameters	89%
axis ratio	89%
Effective aperture diameter	17%
Total magnitudes: Harvard	28%
asymptotic	31%
Surface brightness parameters	17%
Colour indices: B − V	29%
U − B	19%
Radio flux at 1400 MHz	6%
Radio spectral index	11%
Neutral hydrogen flux	11%
Radial velocities: optical	57%
radio	14%

those kinds of data that are most abundant – this was done because of limits of space (page size) and to avoid having too many blank columns. Masses, radii and magnetic fields are known for less than 10% of the stars; mass loss, chromospheric and coronal data exist only for a few percent. One could go on like this, but the conclusion is clear: we do not know much even about the brightest stars.

The situation gets worse for fainter or more distant stars. The largest stellar data base today contains about 6×10^5 stars for which some fundamental parameters (but not all) are known. Since our galaxy is supposed to contain 10^{11} stars, the fraction we are sampling is 10^{-5} of the total!

Similar remarks on incompleteness can be made with regard to the sample of brightest galaxies. Let us use for this the *Second reference catalogue of bright galaxies* by de Vaucouleurs *et al.* (1976). (A third edition is in preparation.) The authors have provided statistical data on the different entries in their catalogue, of which I reproduce part in Table 8.11. I have rounded off the percentages.

As can be seen, the data are rather scanty, except for morphological types, isophotal diameters and radial velocity. Although the next edition will surely close many gaps, let us recall that the SIMBAD data base already contains 6×10^4 galaxies. It is clear that, for the 5.6×10^4 galaxies other than the bright ones the data are much more incomplete. As with stars, we lack data even for many of the brighter objects.

According to cosmologists, the Universe contains at least 3×10^9 galaxies but we have only 6×10^4 in the data bases and basic data for 5×10^3 of the total. Our sample constitutes again only a tiny fraction of the total: 10^{-6}! Thus we 'know' in detail about only a tiny fraction of the Universe and this situation is not going to change dramatically in the next decades.

With all the figures assembled we may go back to the initial question, namely the probable size of data banks. Three conclusions seem to follow rather easily from this chapter. The first one concerns the number of objects. Although this depends very much on the subject, numbers of the order of 10^3–10^5 seem reasonable for the majority of the cases. In the second place, the information available on each object cannot be compressed in most of the cases into a few figures. All subjects tend to grow in breadth with time, in part also due to the fact that we have to keep the old measurements. Thirdly, data tend to grow almost exponentially in many areas, with doubling times of the order of 5–10 years in many cases. Data bases are thus large and keep happily growing. Although such numbers of data are relatively small for any computer, their assembly and updating presents a very considerable investment of time and effort.

Notes on Chapter 8

The first part of this chapter is based upon the paper by Jaschek (1978), suitably updated.

1. Another study on the subject can be found in a paper by Teleki (1985), *Growth of the knowledge of stellar positions*. He also finds a logistic curve for the number of catalogues as a function of time. The same is found by Davis Philip and Perry (1985) for the bibliography of Strömgren and Hβ type photometry.

   The logistic curve of growth is used in many fields of knowledge; for a general introduction see the books by de Solla Price (1963, 1975). Its use constitutes a definitive advance over the old notion of indefinite exponential growth.

2. The section on the completeness of our knowledge is rather instructive; curiously, the author was unable to find this subject examined in the literature currently available. This may be either a lack of interest or an (unconscious) resistance against an admission that our knowledge of the Universe is restricted to a very small (10^{-5}) sample of it. Obviously, few cosmologists with theories about the Universe as a whole like to be reminded of such a fact. The idea that knowledge of but a small fraction of the Universe permits us to draw conclusions about the

whole is based on the (unproven) assumption that the Universe is homogeneous, which is often called the principle of homogeneity. Such a nice name is only a disguise for our ignorance, which, when called 'principle', seems to acquire a status of respectability that the word assumption apparently does not have.

References

Aitken, R. G. (1932) *New general catalog of double stars within 120° of the North Pole*, Carnegie Institution, Washington

Baize, P. (1950) *J. Obs.* **33**, 1

Barbieri, C., Cappacioli, M., Custiani, S., Nardo, G., and Omizzolo, A. (1983) *Mem. Soc. Astron. Ital.* **54**, 601

Batten, A. H. (1967) *Publ. Dominion Astrophysical Obs. Victoria*, **13**, 119

Blanco, V. M. *et al.* (1968) *Publ. US Naval Obs., second series* **21**, 3

Bos, W. H. van den (1926) *Bull. Astr. Inst. Netherlands* **3**, 149

Burbidge, E. M. (1967) *Ann. Rev. Astron. Astrophys.* **5**, 399

Burbidge, G. R., Crowne, A. H. and Smith, H. E. (1977) *Ap. J. Suppl.* **33**, 113

Campbell, W. W. (1910) *Lick Obs. Bull.* **6**, 17

Davis Philip, A. G. and Perry, C. L. (1985) *Vistas in Astronomy* **22**, 279

Finsen, W. S. and Worley, C. E. (1970) *Rep. Obs. Johannesburg, Circ.*, **7**, 203

Harwit, M. (1981) *Cosmic discovery: the search, scope and heritage of astronomy*, Basic Books (New York)

Hauck, B. (1985) *Bull. Abastumani* **59**, 47

Hewish, A. (1970) In *Highlights of astronomy*, **2**. de Jager, C. (ed.), Reidel

Hewitt, A. and Burbidge, G. (1980) *Ap. J. Suppl.* **43**, 57

Hoffleit, D. (1962) *The bright star catalog, third edition*, Yale University Observatory

Hoffleit, D. and Jaschek, C. (1982) *The bright star catalog, fourth edition*, Yale University Observatory

Hoffmeister, C. (1970) *Veraenderliche Sterne*, Barth Verlag, Leipzig

Innes, P. T. A. (1927) *Southern double star catalogue from −19° to −90°*, Union Observatory, Johannesburg

Jahreiss, H. (1985) ESA SP-234, p. 1899

Jaschek, C. (1978) *QJRAS* **19**, 269

Jeffers, H. M., van den Bos, W. H. and Greeby, F. M. (1963) *Index catalog of visual double stars*, Lick Observatory Publications, **21**

Jenkins, L. F. (1952) *General catalog of trigonometric stellar parallaxes*, Yale University Observatory

Jenkins, L. F. (1963) *Supplement to the general catalog of trigonometric stellar parallaxes*, Yale University Observatory

Johnson, H. L. and Morgan, W. W. (1953) *Ap. J.* **117**, 313

Kholopov, P. N. (1985) (ed.) *General catalogue of variable stars, fourth edition*. Vol. I (Andromeda–Crux), Moscow, Nauka Publ. House

Kukarkin, B. P. and Parenago, P. P. (1948) *General catalogue of variable stars* (Moscow, Academy of Sciences of the USSR)

Kukarkin, B. P. *et al.* (1958) *General catalogue of variable stars, second edition* (Moscow, Academy of Sciences of the USSR)

Kukarkin, B. P. *et al.* (1970) *General catalogue of variable stars, third edition* (Moscow, Academy of Sciences of the USSR)

Manchester, R. N. and Taylor, J. H. (1977) *Pulsars*, Freeman

Manchester, R. N. and Taylor, J. H. (1981) *AJ* **86**, 1953

Mermilliod, J. C. (1977) *BICDS* **12**, 53

Mermilliod, J. C. (1984) *BICDS* **26**, 3

Moore, J. H. (1924) *Lick Obs. Bull.* **11**, 166

Moore, J. H. and Neubauer, F. J. (1945) *Lick Obs. Bull.* **521**

Schlesinger, F. and Jenkins, L. F. (1935) *General catalog of stellar parallaxes, second edition* (Yale University Observatory)

Schmidt, M. (1963) *Nature* **197**, 1040

Schnauder, G. (1923) *Ergebnisse der exakten Naturwissenschaften*, **2**, 19, Springer

Solla Price, D. de (1963) *Little science, big science*, Columbia University Press

Solla Price, D. de (1975) *Science since Babylon*, Yale University Press

Teleki, G. (1985) *Bull. Obs. Belgrade* **135**, 40

Vaucouleurs, G. de, Vaucouleurs, A. de and Corwin, H. G. (1976) *Second reference catalog of bright galaxies*, University of Texas Press

Veny, J. B. de, Osborn, W. H. and Janes, K. (1971) *PASP* **83**, 611

Veron-Cetty, M. P. and Veron, P. (1984) ESO Sci. Rep. No. 1

Veron-Cetty, M. P. and Veron, P. (1985) ESO Sci. Rep. No. 4

Wielen, R., Jahreiss, H. and Kruger, R. (1983) In *The nearby stars and the stellar luminosity function*, Davis Philip, A. G. and Upgren, A. R. (ed.), Contr. van Vleck Obs., No. 1, p. 163

Worley, C. E. (1963) *Publ. US Naval Obs., 2nd ser.* **18**, 3

Worley, C. E. (1976) In *IAU Coll. 35, Compilation, critical evaluation and distribution of stellar data*, Jaschek, C., and Wilkins, G. A. (ed.), Reidel, p. 179

Worley, C. E. and Heintz, W. D. (1983) *Publ. US Naval Obs., 2nd ser.* **24**, part VII

Worley, C. E. and Douglass, G. G. (1984) *The Washington double star catalogue*, US Naval Observatory

9

Data banks and data bases

We shall continue in this chapter the matters raised in Chapter 7; there we considered how cataloguing has evolved from private files into teamwork, under the impact of the ever-growing data flux, examined in Chapter 8. We shall look at the situation in more detail and see along which lines data collection has evolved further.

9.1   Storage of information

The best start is perhaps to consider one of the basic steps towards any data collection – namely, the storage of the information. Until very recently, the only way of storing data files was on handwritten cards or sheets. Such a procedure is fine until the moment one wants to publish the data stored, because one must then transcribe the contents in a uniform format and have it typed or typeset. These operations are certainly lengthy and one must still add the proofreading of the pages or the galleys. Because of this, the transcription of a card-file into typescript usually takes one or two years for any moderately-sized catalogue.

Handwritten cards also have the other inconvenience that they are ordered in a certain way – for instance, by right ascension of the object. Any sorting out by another parameter is practically impossible because it would require either re-shuffling the whole original set of cards or doing this operation on a copy of the catalogue. If one remembers that copying machines (for instance of the Xerox type) became available only in the nineteen-fifties, clearly such an operation was out of the question until recently.

A third disadvantage comes from the fact that the contents of cards are rarely fully documented because cards are the work of one astronomer, who remembers what he did; but nobody else knows what he did and so usually the file is not self-explanatory.

The advent of computers brought a number of changes. First of all, computers require a strictly uniform presentation for the input of data, which has the advantage that eventually one single system is used and the data file is understandable to everybody.

Once the information has been entered, it can be listed by any one type of the data; in other words, the file can be re-arranged, for instance by declination or magnitude, without any difficulty. One can also constitute samples defined according to certain criteria (stars south of −30°, between 12 h and 23 h and brighter than 7.0 mag), something that is exceedingly difficult to do with handwritten cards.

Computer read-outs can be obtained immediately and the delays associated with the transcription and proofreading necessary for catalogues based upon hand written cards are avoided.

Finally, a number of checking or cross-checking procedures can be made automatically by the computer and star positions can be precessed from one equinox to another. (When no computers were available, this was a very time-consuming procedure, which was rarely undertaken.)

The first computers were produced in the early nineteenth century but their use for science was not popular at first, one of the reasons being that, except in astronomy, lengthy calculations were usually not needed. Astronomers were the first scientists to use mechanical calculators, which appeared in the late nineteenth century. The use of electromechanical machines (punched card machines) was investigated as early as 1928 by Comrie (1928) but the loan of such a machine was considered to be too expensive when compared to the price of the existing hand calculators. The punched card machines were later used by American astronomers in the nineteen-thirties for celestial mechanics calculations and for astrometric catalogues; a summary of their use is given in a book by Eckert (1940). That same year he became the director of the Nautical Almanac Office of the US Naval Observatory and started introducing punched card equipment for the preparation of the almanacs.[1]

Other Almanac offices followed later on but the conversion to electromechanical machines was only completed in the late nineteen-fifties.

In 1947 transistors were invented, which subsequently led to the construction of electronic computers in the nineteen-fifties. Although they were very expensive at first, their cost-efficiency ratio diminished steadily over the years and in the nineteen-sixties they became accessible even within the budgets of astronomical institutes. The introduction of computers implied a very big change for astronomers and they discarded completely the use of logarithmic and other mathematical tables, which at that time were found on every astronomer's desk (and which can still be seen in large quantities in many observatory libraries).

Besides the advance in the purely computational aspect, which improved constantly over the years, there was a second aspect, namely the memory of the computer, i.e. the quantity of information that could be stored for further use. In the nineteen-forties, the memory was of 10^2 bytes, in the fifties of 10^5 and in the sixties it had surpassed 10^7. With giant memories available, the computer rapidly became the ideal way to handle data collections, which we shall call 'data banks'. We may use the general definition that a 'data bank' is 'a set of coherent information collected for further use'. This definition is rather wide, since it includes also hand-written card files. So we complete it by adding the words 'by means of computers'. This is still rather general because it includes files not yet made computer-readable and so we must add 'and prepared in a computer-readable way'. I have purposely started with an incomplete definition to show that the idea of a 'data bank' is really more general than a computer-readable data bank or data base. We shall use this definition in the rest of the book; please observe, however, that data bank and data base are sometimes used in the literature as synonyms.

It is clear that data banks became computerized because the technology became available; computerization is thus a technical aspect, analogous to the step from handwritten to printed books. It does not change the basic operations that must be carried out to establish a data collection, although it adds a series of operations, specific to the new technique.

9.2　The establishment of a data bank

Let us describe next the basic operations to establish a data bank and consider afterwards which additional steps are required by its computerization.

9.2.1　*Data collection*

The first step is the data collection. We have already seen in Chapter 5, when talking of the presentation of data, that:

(a) the purpose of the collection must be carefully defined;
(b) that data must be ordered;
(c) that data must be explained.

We shall not come back in detail on these points, although it is good to remember that each point presents a host of problems connected with it. To name just a few, we start with point (a). As an example, let us assume that one wishes to produce a data bank on the masses of galaxies. This is thus the purpose and defines the content of the data bank. The first step is to look at the different methods used for the determination of masses, such as rotation curves, pairs of galaxies and galaxies in clusters. Each one of these methods produces data that can be entered in the data bank but, obviously, it is important to know which was the method used to derive a given datum and who did it. This is a rather obvious point but one could quote data bases where it is not done. Necessarily, each method requires additional and perhaps differing information. In pairs of galaxies one must know which is the companion, at what distance it is found and, if a mass ratio is determined, how it was done and upon what observational material the determination was based and what statistical corrections were applied. Also, the morphological galaxy types must be given and the brightness of each component. All this is essential information.

In masses derived from rotation curves the additional information is different. Here one is interested in the observational data used for establishing the rotation curve, the greatest distance from the centre at which observations were made, the resolution of the spectrograph and so on. What model was applied to convert rotation into mass? In clusters we have again other additional data such as the designation of the cluster the galaxy belongs to, the types of cluster members, the radial velocities, and so on.

It is clear from this simple listing that the task of establishing a data bank is essentially the work of a specialist, who has previously worked on mass determination of galaxies. As a rule, the quality of a data bank depends on the professional qualification of its author, and one should never take seriously data banks established by outsiders.

With regard to point (b) we find essentially all the niceties of object designations, dealt with in Chapter 6, plus some additional ones, which can be illustrated by an example drawn from the French data base 'Pascal'. Bibliographers thought it a good idea to introduce an index of objects. This means that one takes the objects quoted in the title and constructs an index with these designations. This very good idea runs into the general diffi- culties shown by all designations. Assume that one author quotes a star he calls α CMa; it will appear as such in the 'index of objects'. Now, if in a second paper the same star is quoted as HR 2392, what is to be done? The bibliographers' answer is 'quote both', which is correct in the sense that one may thus locate both papers but is wrong if one looks for HR 2392 without knowing that it is also called α CMa. As a solution one might quote both references thus

$$\alpha \, \text{CMA} = \text{HR } 2392$$
$$\text{HR } 2392 = \alpha \, \text{CMA}$$

but with up to 35 identifications per star (in some cases), this is a very cumbersome procedure.

Obviously, the solution is the consultation of a comprehensive cross index of designations, like the one of SIMBAD.

As one can see from this example, the ordering by designation can become a complicated problem, for which specific solutions adapted to each case must be sought.

With regard to point (c) one should say that if a data bank has an incomplete description of its contents, or if it does not carry references to the papers the information came from, the data bank is useless, since one cannot check the origin of the information, the quality of the data and the correctness of its transcription.

With all this in mind, assume that one has produced a data bank of mass determinations of galaxies. This is usually the point where the specialist considers his work to be finished. He has a data bank, which he still might convert into a data base, by making it computer-readable, and which contains all usable information on the subject for the moment. Usually the specialist has also made some provision for regular updating, so that the data bank or data base is a really useful tool for him.

Such a data bank might be published without further delay. In the terminology of Chapter 5, it would be called a 'compilation catalogue' and more specifically a 'bibliographic compilation catalogue'.

However, data banks are not published very often and this is due mostly to the purpose for which it was constructed. If the data bank was considered as an essential step for the constructors own research, he will probably not wish to publish it until he has used it fully. Since the possession of an up-to-date data bank is essential for much general statistical research, he will naturally hesitate to publish it because, by publishing it, he opens it to all competitors in the field who may use it immediately for essentially the same purposes the constructor of the data base had himself.

Instead of the three or four papers he could expect to publish, he gets credit for only one (i.e. the catalogue), which is usually not regarded as 'true' research. On the other hand, if he publishes it when he has finished using it, the data base will probably be out-dated, and not be worth publishing. Whether to publish a data base or not is sometimes a hard decision to make. It might be observed that most objections against publishing are exaggerated – in every case the author knows the data base much better than anyone else and can use it much more efficiently than any competitor.

In general one finds that, whereas data banks created by one individual astronomer often remain unpublished, groups usually (but not always!) tend to have a more positive attitude towards publication, probably because the data bank is the result of teamwork.

Let us assume that the producer(s) want to publish the data bank or base. One possibility would be to leave it at the level of a 'bibliographic compilation catalogue' as mentioned above, but it would also be possible to proceed further.

9.2.2   *Critical evaluation of data*

If more work on the data base could be done to ascertain the quality of the data or to derive 'best values', one could create a 'critical evaluation of data' and the outcome would be a 'critical compilation catalogue'. The work necessary to evaluate data again has to be done by a specialist in the field of the data collected, and in some sense it is the specialist's essential role. In a world in which science is split into ever-increasing numbers of special fields (and subfields) the community as a whole has to use the work of *the* specialist in a given field. It is thus the privilege and the duty of the specialist to prepare the data in such a way that they can be integrated into the whole fabric of astronomy for further use.[2] The steps to be followed for critical evaluation can be illustrated with a case taken from stellar photometry.

Assume that a data bank exists for UBV photometry of stars. We have N_1 measurements referring to N_2 stars, coming from N_3 different publications. Since authors normally indicate the errors affecting the measured colours and magnitudes, each measurement can be regarded as a data point in a three dimensional continuum with axes V, U − B and B − V. If no systematic errors were present, the measurements corresponding to the same star should coalesce into one data point, or at least lie as close together as errors indicate. One may imagine each point surrounded by a 'probabilistic' ellipsoid whose axes are equal to the errors in each coordinate.

However, in practice, one finds sources of inhomogeneity (Hauck 1973, Nicolet and Hauck 1977) due to:

(a) the photomultipliers
(b) the filters
(c) the methods of reduction to outside the atmosphere
(d) the lack of or the bad use of standards
(e) the optical properties of the telescope
(f) the cooling of the photomultiplier.

The photomultipliers and filters used have small differences from the average properties indicated by the manufacturer, which appear later on as small systematic errors. Similarly, the incomplete modelling of atmospheric extinction can lead to more systematic errors and the lack of or bad use of standard stars to further systematic residuals. Other systematic effects may be introduced by optical elements (lenses) or by unclean or dust-covered surfaces, and/or by lack of or imperfect cooling of the photomultiplier. For a general description of errors plaguing photometry see Rufener (1982). It is therefore dangerous to take simple averages of all published measurements of a star without correcting first for possible systematic errors.

The first step is to select a publication (or a set of publications) as reference. For the UBV system one may take as such the list of observations by Johnson, who introduced the system (Johnson and Morgan, 1953). From this list all variable stars, and stars suspected of being variable, are eliminated.

One then compares each list (i.e. each one of the N_3 publications) with the chosen reference list and retains for comparison all stars common to both lists. Let us call c_i(list) and c_i(reference) the values of c_i for the same star in both lists. We use c_i as an abbreviation for $c_1 = V$, $c_2 = U - B$ and $c_3 = B - V$. We write then

$$c_i(\text{ref}) = a + b \cdot c_i(\text{list}) + d \cdot c_1(\text{list}) + e \cdot c_2(\text{list}) + f \cdot c_3(\text{list}).$$

We expect $a \simeq 0$, $b \simeq 1$ and d, e and f to be very small. Regression lines, established for each parameter, permit us to derive the average deviation (ΔC) and the standard deviation (σ) that characterize the specific comparison of a given list with the reference list.

An examination of the ΔC and/or σ will allow one to:

(a) eliminate the list,

(b) derive systematic corrections, or

(c) keep the measurements as they are published.

In cases *b* and *c* the comparison of the list with the reference permits us to attribute a weight w_i to list *i*.

One difficulty often encountered is that the overlap between the list and the reference is empty or contains too few stars to be significant. If this is the case, one has to start enlarging the 'first reference' list by successive additions of those lists which present the largest number of stars in common.

The process is finally completed by deriving an average colour or magnitude for each star from a combination of the corrected parameters of list *i* into

$$c = \frac{\Sigma \, c_i w_i n_i}{\Sigma \, w_i n_i}.$$

n_i is the number of measurements that went into the observation given in a list. One may also compute σ, which measures the deviation of each observed (and corrected for systematic errors) parameter from the average. If this value is larger than expected, one must examine whether the star is variable or double or was misidentified or whether there has been a clerical error somewhere. This is an important, but also a lengthy, tedious and delicate task. (See, for instance, Mermilliod (1988a).)

We have described in some detail this procedure because it is a rather clear-cut case. Other cases, for example when one is calibrating observed quantities with a theoretical parameter, may be much more complicated. Take as an example the establishment of effective temperatures from observed photometric parameters (colours of stars).

Here the crucial problem is the establishment of a good list of stars having well-known effective temperatures. One then correlates critically evaluated photoelectric colours with them and obtains a tabular relation between effective temperatures and a given colour system. (See, for instance, Hauck (1985).) Upon closer examination, a number of difficulties rapidly appear, which are connected mostly with the establishment of the list of stars with well-known effective temperatures. To derive them we need to know the total flux, the distance and the radius of the star. Of these quantities the total flux is only known if one can observe the flux in *all* wavelength regions, which is rarely the case. The radius is only directly measurable in a few nearby stars and in eclipsing binaries. With regard to the distance the situation is even worse, since errors of less than 5% are only attainable for stars nearer than 15 or 20 pc.

All this can be summarized by saying that to obtain a list of effective temperatures one needs as the first step critically evaluated lists of three other parameters. This case illustrates well the point that for further progress one needs critically evaluated data of *all* kinds. Often such data are used afterwards for applications very different from those foreseen when they were derived.

From the examples given it seems obvious that the number of comparisons and calculations to be made is so large that they can only be done by a computer. This ease of comparison is one additional factor that has accelerated the conversion of data banks into data bases.

We have mentioned before that critical evaluation is the next logical step in the development of a data bank. Because the number of data increases rapidly (see Chapter 8) it is clear that critically evaluated data are also transitory because, when large additions to the number of data occur, a new evaluation is required. On the other hand, given the complexity and the ensuing amount of work they generate, critical evaluations are not carried out very often. In practice, data are collected (and eventually published) fairly frequently, but critical evaluations are published only when very large bodies of data or new techniques demand it. For instance in UBV photometry we had critical evaluations in 1976 and 1986 (Mermilliod and Nicolet 1977) although updates of the compilations were published in 1979 and 1982. On the other hand, in many areas we still have no recent critical compilation, as in stellar radial velocities; the last one, made by Wilson (1953), was partially superseded by that of Evans (1967). In other areas, a critical evaluation has never been attempted: many parameters referring to external galaxies are still in the state of having only (or at most) bibliographical compilation catalogues.

9.3   Requirements introduced by the use of computers

We consider next the requirements imposed on a data bank when the step to convert it into a data base is undertaken. We can summarize (Davis, 1977) these under four headings, namely:

    (a) security
    (b) resilience
    (c) standards
    (d) compatibility.

9.3.1 *Security*

One may define security as a protection of the data base against unauthorized access and changes in data. The term thus implies an 'external' security, which is also called 'privacy', and an 'internal' security, also called 'integrity'.

In data banks the security is provided by the key that locks the cupboard in which the handwritten cards are stored. In data bases, security is most often assured through the use of passwords, also called 'keys' or 'locks'. Although security is becoming a central problem for data bases with economic value, in astronomy, the problem seems not to have needed (up to now!) serious attention and users of more sophisticated security systems would probably be horrified by the lack of precautions. Much more attention has been paid to integrity, which is the protection against corruption of data. We have seen that the first step in a data collection is the documentation, namely the explanation of the computer files and its headers and the computer program for using and printing the files (like code, density and block size).

Nevertheless, documentation is not the only aspect of integrity because we also have the problem of the accuracy with which the machine-readable data were transcribed from the original sources. The first check is thus 'proofreading'. According to the experience at the CDS, the number of transcription errors diminishes when the operator can visualize on the screen what he has punched, as compared to the case when he does not see the result (as was the case with punched cards). An error rate of 5 per thousand is a good performance but often errors of transcription are much more frequent (Wagner 1986); the only protection is proofreading. Usually one either proofreads on a screen or on a print-out, or one compares two versions (transcribed by different persons) of the same set of data. This proofreading is completed with some tests of correctness introduced in the program, such as checks for alphabetical or numerical order ($n_2 < n_1$ for example, or no 'Bright Star catalogue' number greater than 9110); of zeros and blanks; of parameter ranges (no magnitudes outside the interval (-2 to 30) and validity tests. Unfortunately, the integrity is not guaranteed by proofreading alone, because the original data themselves may be in error, as we have discussed in Chapter 5. Such errors become apparent when the data are compared with other data of the same type – for instance, when collecting all UBV photometric data – or with other data (for instance an F5V star having $U - B = -0.30$) or simply by astronomers who examine data in detail. By experience, most such errors come from insufficient proofreading of the original paper, although more perverse cases occur from time to time and there are disturbing tales of errors that have persisted for over one century. Probably, part of the proofreading problem will disappear in the near future when optical readers, i.e. devices that scan, read and memorize (on disc or on tape) a printed text. Although attempts up to now have not been completely satisfactory, this is clearly only a question of time.

9.3.2   *Resilience*

This is the capacity of a system to recover from errors of system, program or hardware types. Or, in other words, it is the guarantee against destruction or corruption. It is analogous to the problem of observational plate archives in which the plates must be safeguarded against degradation, humidity, temperature excess, sunlight, fire, theft, breaking and so on.

The simplest safeguard is to keep copies at different places. Notice, however, that spontaneous corruption of data stored on tape may occur if the tape is not used, say, for more than a year. One must thus re-run tapes at intervals of the order of a year to make sure that no spontaneous corruption occurs. Probably this problem will also disappear in the near future with optical numerical discs. (Notice, however, what we said on page 38, namely that the lifetime of such devices is as yet unknown.) Re-runs are in general satisfactory to safeguard existing files, but they may not be sufficient when one is working on a file, for instance updating it. It can then be advantageous to dump the files periodically, an operation called back-up (or journalization). When one updates, one creates a new updated 'son' file. When this is deemed to be correct, it replaces the previous file – called the 'father' file – and the 'father' file replaces the 'grandfather' file, which is then destroyed. This is equivalent to keeping the current file *and* the previously updated file (the father file) in case an accident happens.

One should be aware of the fact that such accidents inevitably happen, despite the precautions one takes; journalization is thus a very useful practice.

Let us add that the updating of files (or in general any change in the files) is an activity reserved for the data base administration – the user cannot be allowed to modify data in data banks. Measures must thus be taken to prevent him from doing so, and we shall come back to this problem in the next chapter on data centres.

9.3.3 *Standards*

We have already seen that a very large number of catalogues and data bases exist. As long as they were few, each producer could follow his own preferences in matters of presentation and the user had to adapt himself to the different situations. If the number of data bases becomes large, everything pushes toward standardization, to facilitate the use of catalogues. Such standardization should refer to various different things, namely:

    (a) use of standard units, and adherence to IAU conventions and notations (see Chapter 3);
    (b) rigorous checking to ensure conformity with written sources (see above, integrity);
    (c) source reference and authorship of both source and tape producers;
    (d) description of content (headers and trailers);
    (e) presentation of data.

We may omit details of points *a* and *b*, which have already been discussed. Point *c* is also clear because it testifies to the respect of intellectual property. Point *d* is also reasonably clear and one feels perhaps that it should not even be mentioned; it is a fact of life, however, that specialists, being well acquainted with their data, are rather careless about what they consider to be 'obvious', which usually happens not to be obvious at all to outsiders.

The last point (presentation) becomes critical when the content of the data base is circulated among users, usually in the form of a tape. It is then advisable to follow certain rules, which are discussed in detail by Hill (1981), Warren (1985), Warren (1986). These authors recommend a certain number of technical points and a standard of presentation. Among the technical points, it is worthwhile retaining the following:

    (a) data sets should contain a simple logical record per data catalogue entry (i.e. no multirecord format);
    (b) data should conform to a format which is conducive to true machine processing and, more specifically, data files should be readable with a simple Fortran 'read' statement;
    (c) the logical record length should be great enough to accommodate all data fields without special treatment.

A few more points refer to the transcription of notes and of spectral types and other parameters; the reader is referred to the three papers quoted for full detail.

The standard of presentation proposed comprises five points, namely:

(a) introduction and source reference;

(b) tape content – this provides the header, with a byte-by-byte format description;

(c) tape characteristics – this provides indigenous parameters like number of files, logical record length, record format and total number of logical records for each file;

(d) remarks, modifications, acknowledgements and references to errata and to bibliography – this is the so-called trailer;

(e) sample listing.

A (printed) document containing these five items is distributed with each magnetic tape. Such a standard documentation of catalogues is very useful because it makes it clear exactly what is on the tape. The drawback is that it is a written document, which has a tendency to go astray at the moments one needs it.

The logical conclusion is to write things on the tape itself (except of course the specifications needed to read the tape). This is already done by some authors, for instance in photometric catalogues established by the Lausanne-Geneva group. And the last step would be to standardize the specifications of the catalogue and its arrangement, with, for instance, ASCII tapes, density 1600 bpi with blocks of 2880 ($= 36 \times 80$) bytes. This has been proposed by Ochsenbein (1986). In short, it is similar to the FITS standard introduced for the circulation of two-dimensional images (Wells *et al.* 1981), extended later to tables (Harten *et al.* 1984). Regretfully, no international convention could be agreed upon until now.

9.3.4   *Compatibility*

The last problem we have to examine is the one of compatibility. In some ways this is becoming a problem of the past – it is a characteristic problem whenever a new industry develops. At the start each manufacturer uses its own standards, until the laws of the market and the number of users force manufacturers to agree to a very few or even a single standard. The physical system of units, the sensitivity of photographic material, the specifications for the number of grooves in gramophone (phonographic) records, are now all reduced to a very few standard systems, after an initial period of anarchy with dozens of standards.

Exactly the same happened with computers at the beginning – each manufacturer produced its own standards and for a decade or so, many computer types were incompatible in the sense that they could not read what other computers had produced.

This is now practically overcome and all computers can read the tapes. Some problems still remain with discs and similar devices, but hopefully these will also disappear within a short time, again through competition between manufacturers.

This completes my description of data banks and data bases. I shall examine in the next chapter the evolution that followed. As a last point I will mention briefly the existence of a directory of data bases in astronomy (Jaschek 1980) in Codata Bulletin No. 36. A second edition of it will be published in 1988. The first edition listed about 75 data bases. Such a number seems large, but according to Behrens and Ebel (1982) there are three thousand data compilations in physics (!).

For further information on data bases, see also the section CODATA in Chapter 13.

Notes on Chapter 9

For a general introduction on data banks and data bases, see Eliezer (1980). For the more specialized problems in physics, see Shaw (1985). In astronomy the best survey is given by Hauck and Sedmak (1985).

1. For the introduction of computers in science, see for instance Moreau (1982). No general history of computing in astronomy exists, to the author's knowledge. The best is one excellent paper on computing in celestial mechanics by Seidelmann (1984), which provides also references to other sources.

   The development outside the US was considerably slower. Prof. T. Lederle, in a private communication, provided some data on the evolution of computing devices at the Astronomisches Rechen-Institut. Although the first use of punched card machines in Germany was described by Peters (1935), a result of the Second World War was that these machines were not used until the nineteen-fifties (Lederle 1956, Fricke 1962), i.e. considerably later than in the US. See also Pollack (1928).

2. For the critical evaluation of data, the reader is referred to various papers in IAU Colloquium 35 (Jaschek and Wilkins 1977) and, more specifically, to the paper by A. Underhill, J. M. Mead and T. A. Nagy. The international Colloquium of Tbilisi (Kharadze 1985) contains another series of important papers.

Another useful reference is IAU Symposium 111 (Hayes, Pasinetti and Davis Philip 1985).

A first list of data bases is published in *BICDS* No. 33; this list and others will be summarized in the second edition of the *CODATA Directory*.

A large literature exists on the construction and use of data bases; see, for instance, Ullmann (1980). An excellent review of the problems connected with scientific data bases and with astronomical ones in particular, is given by Murtagh (1988). He also provides many references to the recent literature on the subject.

References

Behrens, H. and Ebel, G. (1982) *CODATA Bulletin* **48**

Comrie, L. J. (1928) *Vierteljahresbericht Astr. Ges.* **63**, 317

Davis, M. S. (1977) In *IAU Coll. 35, Compilation, critical evaluation and distribution of stellar data*, Jaschek, C., and Wilkins, G. A. (ed.) Reidel

Eckert, W. J. (1940) *Punched-card methods in scientific computation*, New York, Columbia University

Eliezer, L. (1980) In *Data handling for science and technology*, Rossmassler, S. A. and Watson, D. G. (ed.) North Holland

Evans, D. S. (1967) In *IAU Symp. 30, Determinations of radial velocities and their applications*, Batten, A. H. and Heard, J. F. (ed.) Academic Press, New York

Fricke, I. J. (1962) *Astr. Rechen Institut Heidelberg, Mitt. Ser. A* No. 16

Harten, R. H., Grosbol, P., Tritton, K. P., Greisen, E. W. and Wells, D. C. (1984) *Astron. Image Processing Circ.* **10**, 14

Hauck, B. (1973) In *Spectral classification and multicolor photometry*, Westerlund, B. and Fehrenbach, Ch. (ed.) Reidel, p. 285

Hauck, B. (1985) In *Calibration of fundamental stellar quantities*, Hayes, D. S., Pasinetti, L. E. and Davis Philip, A. G. (ed.) Reidel, p. 271

Hauck, B. and Sedmak, G. (ed.) (1985) 'Data handling in astronomy and astrophysics', *Mem. Soc. Astron. Ital.* **56**, Nos. 2 and 3

Hayes, D. S., Pasinetti, L. E. and Davis Philip, A. G. (ed.) (1985) *IAU Symp. 111, Calibration of fundamental stellar quantities*, Reidel

Hill, R. S. (1981) *Astronomical Data Center Bulletin* **1**, 69

Jaschek, C. and Wilkins, G. A. (ed.) (1977) *IAU Coll. 35, Compilation, critical evaluation and distribution of stellar data*, Reidel

Jaschek, C. (1980) *CODATA Bulletin*, **36**

Johnson, H. L. and Morgan, W. W. (1953) *Ap. J.* **114**, 522

Kharadze, E. D. (ed.) (1985) 'Stellar catalogues: data compilation, analysis, scientific results', *Abastumani Bulletin* **59**

Lederle, T. (1956) *Nachrichtentechnische Fachberichte, Braunschweig*, vol. 4, Anlage 2

Mermilliod, J. C. (1988a) *AA Suppl.* in press

Mermilliod, J. C. (1988b) *AA Suppl.* in press

Mermilliod, J. C. and Nicolet, B. (1977) *AA Suppl.* **29**, 255

Moreau, R. (1982) *Ainsi naquit l'informatique*, Dunod

Murtagh, F. (1988) *BICDS* **34**

Nicolet, B. and Hauck, B. (1977) In *IAU Coll. 35, Compilation, critical evaluation and distribution of stellar data*, Jaschek, C. and Wilkins, A. G. (ed.), Reidel, p. 121

Ochsenbein, F. (1986) In *Data analysis in astronomy II*, Gesu, V. di, Scarsi, L., Crane, P., Friedman, G. H. and Levialdi, S. (ed.), Plenum Press, p. 363

Peters, J. (1935) *Vierteljahresschr. Astron. Gesellschaft* **70**, 314

Pollack L. W. (1928) *AN* **253**, 91

Rufener, F. (1982) *BICDS* **22**, 2

Seidelmann, P. K. (1984) In the article 'Celestial mechanics', *Encyclopedia of computer science and technology*, Vol. IV, Decker, New York and Basel, p. 243

Shaw, D. F. (1985) In *Information sources in physics*, Shaw, D. F. (ed.), Butterworth

Ullmann, J. D. (1980) *Principles of data base systems*, Computer Science Press

Underhill, A., Mead, J. M. and Nagy, J. A. (1977) In *IAU Coll. 35, Compilation, critical evaluation and distribution of stellar data*, Jaschek, C. and Wilkins, A. G. (ed.), Reidel, p. 105

Wagner, M. J. (1986) *BICDS* **30**, 107

Warren, W. (1985) *Mem. Soc. Astron. Ital.* **56**, 285

Wells, D. C., Greisen, E. W. and Harten, R. H. (1981) *AA Suppl.* **44**, 363

Wilson, R. E. (1953) *General catalogue of stellar radial velocities*, Washington, Carnegie Institution Publ. No. 601

10

Data centres

We have seen in the previous chapter how data banks got converted into data bases in the nineteen-sixties, under the impact of computer technology. We shall consider in this chapter the evolution which followed and which brought the creation of data centres (DC). After analysing DCs we shall consider their later evolution and the prospects for the future.

10.1   The establishment of data centres

Chapter 9 explained in some detail how computers permitted the conversion of written catalogues and data files into machine-readable versions. Usually, such versions were initially prepared for the internal use of an observatory. Because of the rapid generation of machine-readable versions of catalogues, some duplication of effort was almost unavoidable.

In 1970 the International Astronomical Union, in collaboration with COSPAR and other Institutes, established the 'International Information Bureau on Astronomical Ephemerides' (BIIEA) at Paris, whose purpose was to provide information on the availability of ephemerides and star catalogues in machine-readable form. Between 1971 and 1976, 125 information cards were distributed. Each information card listed catalogues available on tape so that a user could contact the tape producer and eventually obtain a copy of the tape.

The next innovation was the creation in France of an Institute – the Centre de Données Stellaires at Strasbourg – devoted specifically to the collection, critical evaluation and handling of data. The father of the idea was Jean Delhaye, who was able to convince a number of European colleagues of his idea, among them Fricke, Golay and Blaauw. The new institute (within the Strasbourg Observatory) was created in 1972 and J. Jung nominated his first Director, who was to be succeeded in 1974 by the writer.

The first task envisaged at Strasbourg was to obtain copies of all existing machine-readable catalogues. With such a collection of catalogues it became possible to provide an efficient distribution service. A potential customer no longer needs to search in the Information Cards of the BIIEA, whether and where a catalogue is available – he applies to the CDS, which guarantees that he can get all that is available.

The response of astronomers to such a service was encouraging and in the next decade a number of countries set up similar organizations. We may mention the United States (Astronomical Data Center, NASA, Greenbelt, Maryland), the German Democratic Republic (Zentral-Institüt für Astrophysik, Potsdam), the USSR (Astronomical Data Centre of the Academy of Sciences, Moscow), and Japan (Kanazawa Institute of Technology).

Notice the name used: data centre (DC). We may define a data centre as an institute devoted to the collection, critical evaluation and distribution of data, which is almost exactly how the mission of the CDS was defined at its creation. To be complete, the objectives of the CDS also included 'research done on the data collected'; we shall come back to this later on.

Implicit in this definition is that a DC is not a data producer. This is an essential characteristic; the DC acts only as a distributor of catalogues, not as their producer. The reason is simply that the data bank (or base) producer has to be a specialist and, if possible, an outstanding specialist. A data centre never has expertise in all fields; it acts as a librarian who distributes, but does not himself produce, books (although rare exceptions do exist). The DC astronomers, like librarians, can exert only an indirect influence in trying to discriminate against bad work through a selection of what they admit into their collections.

On the other hand, a DC can encourage production of new catalogues, by appealing to well-known specialists and pointing out what is desirable or what is needed. They can also contribute to the conversion into machine-readable form of information which is not presented this way – for instance, the many catalogues published in print. But in this also the DC astronomer must follow what specialists recommend him to do, because otherwise he might spend valuable time on catalogues that are superseded or unimportant. Since the DCs are interested in obtaining the largest possible number of catalogues in machine-readable form, the easiest way to achieve this is through international collaboration. DCs have thus made bilateral agreements on data exchange, and the system which started with an agreement between the DCs of France and the US has gradually been extended to the other ones, in the Soviet Union, the German Democratic Republic and Japan. Usually such agreements include provisions for data exchanges on a regular basis (once or twice a year), in which new catalogues are exchanged, and provisions for exchanges of specialists (one week per year, typically). Such agreements help to avoid gaps in the coverage and guarantee that astronomers in all DCs share all existing information.

Such a procedure also limits the difficulties of tape transportation. Even in our times, if a tape has to cross a national border it may be delayed because of customs problems and, in Western Europe, mailing tapes usually requires as much time as in the last century when parcels were transported by stage coach. By creating data centres in different countries or geographical regions, astronomers do away with at least some of these difficulties.

Other countries have preferred not to set up their own DC, but simply to have a subcentre – i.e. an observatory that agrees to act as a national distributor for data collected at one data centre and transmitted to it. There is such an arrangement in the UK, where the Rutherford Appleton Laboratory acts as a national distributor for the catalogues it receives from France (CDS), and in China, with the Beijing Observatory. Such an arrangement obviates the need to have a separate British or Chinese organization; at a time when manpower is a major problem in astronomy this is an advantage. It is clear that an active exchange organization reduces the need for many independent DCs and makes the national subcentres more attractive.

10.2   Organizational aspects of a data centre

We consider next the organizational aspects of a DC, and examine successively:

(a) standardization for the provision of catalogues;
(b) announcement of availability;
(c) economic aspects;
(d) manpower problems;
(e) legal aspects.

10.2.1   *Standard procedures*

The handling of a large number of magnetic tapes obviously requires standard procedures, both for writing the tapes and for explaining

Table 10.1. *Areas of interest*

– Section 1: Astrometric data

   Contains catalogues referring to stellar positions, proper motions
   and parallaxes. Includes also visual double star observations,
   speckle-interferometry, occultations.

– Section 2: Photometric data

   Contains photometric data, observed from the ground and/or from
   outer space. Variable stars and interstellar polarization catalogues
   are also included here.

   For radio, X-ray and γ-ray data, and surface photometry see
   section 7.

– Section 3: Spectroscopic data

   Contains catalogues referring to spectral classification, radial and
   rotational velocity, lists of spectroscopic binaries,
   spectrophotometry, abundances, line-identifications in stars,
   equivalent widths.

– Section 4: Cross-identifications

   Contains cross-identification tables.

– Section 5: Combined data

   Contains catalogues providing various kinds of data for the same
   object (like the *Bright star catalog*). Includes space velocities and
   orbits of visual or spectroscopic binaries.

– Section 6: Miscellaneous data

   Contains catalogues falling outside other sections: bibliographic
   catalogues, observation logs, photometric sequences, gf-values,
   multiplet tables and sensitivity functions of photometric systems.

– Section 7: Extended and non-stellar objects

   Contains data referring to non-stellar (extra-solar system) objects.
   Includes clusters and associations considered as a whole.

---

their content. This was discussed in the previous chapter under the heading
'standards'; notice that these procedures were proposed by DC specialists.

At another level there is also a uniform designation of catalogues, in the
sense that the same catalogue carries the same number in all DCs, to
facilitate handling. The numbering is done by CDS and NASA, through an
exchange of telex communications. The numbering system consists of
dividing astronomy (outside the solar system) into the areas of interest
listed in Table 10.1. Catalogues in each area are then numbered in
chronological order of appearance. Each catalogue has a four-digit
number, with the first from the left denoting the area and the other three
the number – thus 1010 is the tenth catalogue of area 1 (astrometry).

The NASA DC uses some more symbols to denote the status of a
catalogue. The system is explained in Table 10.2.

Table 10.2. *Status codes for catalogues*

A – Available for distribution
B – Basically checked out on computer, but documentation not yet completed
　　　or some questions remain
C – Catalogue on hand, but not yet checked out by computer
D – Catalogue in preparation, revision, or update (temporarily unavailable)
E – Catalogue has been requested, but not yet received
F – Available in microfiche version
G – Available in both microfiche and microfilm versions
M – Available in microfilm version
R – Catalogue on hand, but we are not authorized to distribute
T – Full documentation available
U – Separate document unnecessary – adequate description on tape

---

10.2.2 *Availability*

Catalogues are announced regularly by all data centres, updating previous lists. Updates are issued at irregular dates, whenever it is convenient, except in the case of the CDS which publishes them periodically in its *Bulletin d'Information du Centre de Données Stellaires* (abbreviated BICDS). This Bulletin is the main publication in the field of astronomical data activities; it appears twice annually and has run since 1971. The NASA DC also publishes a Bulletin, which appears at regular intervals, and of which four issues have appeared up to the time of writing.

The 'announcements of catalogues' provide title, name of author, current number and items such as status (Table 10.2) or availability, or short descriptions of the catalogue itself. These announcements differ from one DC to another.

Some of the catalogues announced may also be available in forms other than magnetic tape, namely microfiches (see Chapter 11) or in the form of a listing. The former way allows large amounts of information to be distributed easily, since 220 printed pages are contained in *one* microfiche. However an optical device is required to read the microfiche.

Listings or print-outs are not very convenient if the catalogue is long, and are more expensive to transport than the tape.

10.2.3 *Economical aspects*

We shall group under this heading the problems of cost and transportation.

DCs can be run either as central facilities the costs of which are paid for by the communities they serve, or as institutions that provide service for which they charge. The CDS, for instance, charges a small amount for the computer time used to copy the tape, plus the cost of the tape, plus the handling charge. The second and third items together constitute 80% of the final bill charged. Handling costs cover the postage, the cost of the package in which the tape is sent and the percentage banks take out from cheques made out in foreign currency. (Commissions may be quite high – 13% for a ten-dollar cheque.)

The NASA and USSR DCs do not charge anything, because they are a public service, although empty tapes may be requested in exchange. It is clear that the CDS policy is not meant to finance all operational costs, but simply constitutes a safeguard against certain abuses which could conceivably occur when the service is completely free. People tend to be less conscious in such cases that somebody somewhere has to pick up the bill.

10.2.4 *Manpower*

It is clear that the operation of a DC implies that somebody does the work and these 'somebodies' have to be astronomers with a fair knowledge of computing and a good deal of enthusiasm for carrying out their work. The fact that DC users often ask quite complicated questions on the content of the DC clearly implies that they want an astronomer and not only a computer specialist as their contact. The 'good deal of enthusiasm' is needed to overcome the notion that everybody wants to use a DC, but very few want to work in it. Usually such work is regarded as less essential to the progress of science than what is called the 'frontier research'. Given the modern organization of society and of science, in which any advance is made through the contributions of many different individuals, it is a viewpoint as wrong as thinking that physics was made by individuals such as Newton, Einstein and Maxwell alone. DCs were created in the 1970s because the progress of astronomy made them necessary – and thus work at them fulfils a need, and must be considered as an important contribution to astronomy.

The staff of data centres are thus composed of a small number of specialists, usually less than ten. Their number may be completed by computer specialists devoted to the smooth operation of the machines, but here again each DC presents a different case.

10.2.5 *Legal aspects*

Since DCs are distributing products not manufactured by them, they must have permission to do so. This is a fundamental requirement stemming from the nature of scientific research as intellectual property. Since astronomical data have little direct economic value, astronomers generally neglect this aspect. It is, of course, very different in other sciences; think for instance of geophysical data bases on petroleum, or chemical data bases on polymers. The fact that the direct economic value is small does not imply that they do not have an economic value at all; consider, for instance, the star catalogue needed for the orientation system of a satellite. If such a catalogue did not exist, it would have to be established at considerable cost.

It is thus natural that DCs should ask the catalogue producers for permission to distribute their work on a non-profit basis. Usually such permissions are granted without difficulty, if the astronomer or an astronomical magazine or society is the intellectual property owner, but the situation may be quite different if the owner is a commercial publishing company. Sometimes contracts with authors contain a clause that the published data, including tape versions of it, are the property of the publishing company. In such cases, tape versions are unavailable for further distribution by DCs, and are available only for the private use of the buyer. Although such practices are rare, they nevertheless constitute a major hindrance for further use of the data. Future authors should do all that is in their power to prevent such clauses appearing in contracts. This is a reasonable request because publishing houses are supposed to be there for printing astronomical information and not for preventing other uses of it.

A further legal aspect concerns the tape versions themselves. Assume that a catalogue was established by X and a machine readable version of it was produced by Y. Everyone accepts that the name of X should be credited, but often references to the tape producers are omitted. This is an unfair practice, which should be corrected (see also Chapter 5), because the production of a good tape version may represent considerable work.

10.2.6 *Existing data centres*

Having said that much on the organization of DCs, let us conclude this part of the chapter with the addresses of the existing DCs, which are given in Table 10.3, and to which the reader may apply with requests.

Table 10.3. *Data centres*

1. National Space Science Data Center (NSSDC)/ Code 633, NASA.
   Goddard Space Flight Center, Greenbelt, Maryland 20771
   Telephone: (301)344–8310;   TELEX 89675 BASCIL GBLT.

2. Centre de Données Stellaires, 11, rue de l'Université, 67000 Strasbourg –
   France   Telephone: 88–35–82–00   Telex: 890506 Starobs F

3. USSR Academy of Sciences, Astronomical Council USSR (Astrosoviet),
   48, Pyatnitskaya Street, Moscow 109017
   Telephone: 231–5461   Telex: 412623 SCSTP SU

4. Zentralinstitut für Astrophysik, Sternwarte Babelsberg
   Rosa-Luxemburg Str. 17 A, DDR–1502 Potsdam
   Telephone: Potsdam 762321   Telex: 15305 VDE PDM

5. Kanazawa Institute of Technology, Department of Physics,
   Nonoichimachi, Ishikawa 921, Japan
   COMPUTER NETWORK ADDRESS: BITNET: JPNKIT

10.3   Integrated data bases

We have defined DCs as being essentially distributors of machine-readable information. The key to further evolution lies in one of the functions that the CDS was supposed to carry out in addition, namely to do research on the data collected (see p. 127). When work started on galactic structure, at the CDS we ran immediately into the problem of the multiple designations of a given object, which I have analysed in Chapter 6. We have also seen that the solution came through the organization of a cross-identification file (CAOI) which lists all synonyms and permits the retrieval of all designations of a given object. This is of course an open catalogue, in the sense that it is updated regularly, so that a 'definitive' CAOI does not exist. The cross-identification file (at first called the catalogue of stellar identifications – CSI) was initiated by Jung (1971). Progress reports can be found in Ochsenbein *et al*. (1977) and Ochsenbein and Bischoff (1982); the most detailed description is given in Ochsenbein *et al*. (1981).

The main use of this index is that it enables one to search for an object in the various different catalogues under any of its aliases.

10.3.1 *SIMBAD and its applications*

The next step in the development, which followed almost automatically, was the use of the CSI to connect all catalogues. It is very tempting to produce a data base containing all known objects (initially only stars), each one followed by a list of the different parameters known. At first, only positions, magnitudes and spectral types were used but, obviously, this was only an appetizer for more. We may leave aside the various steps, the history of which is given by Egret (1983) and refer only to what exists now, which constitutes a unique data base called SIMBAD. This acronym stands for 'Set of Identifications, Measurements and Bibliographic references of Astronomical Data'.

SIMBAD provides for each object the following four types of information (Ochsenbein 1984):

  (a) Fundamental data
  (b) Identifications
  (c) Observational data
  (d) Bibliographic data.

The 'objects' considered in SIMBAD are stars (about 6×10^5), galaxies (about 6×10^4) and other objects (clusters, clouds, etc., of which there are 6×10^3).

Fundamental data are, in this context, data that are considered more important than others. These are, for stars:

  – equatorial coordinates 1950.0, with an indication of precision
  – proper motion components
  – spectral classification
  – B and V magnitudes.

It may happen, of course, that one of these quantities is missing but this occurs in only 15% of the stars and in 10% of the galaxies included in the data base.

For galaxies the fundamental data are:

  – equatorial coordinates
  – dimensions
  – morphological types
  – integrated B and V magnitudes.

Identifications is essentially a list of all synonyms of the object, up to 35 for a single object, drawn from about 300 different designation lists. These identifications can be used to access other data contained in SIMBAD.

Observational data are other data (radial velocity, proper motion, colours etc.). Each of the data given has a bibliographic reference; when available, both mean and individual values are provided.

Bibliographic data are provided from 1950 on. Non-stellar objects, however, have been included systematically only since 1983 (Dubois *et al.* 1983). They provide references to papers in which the object is quoted, giving authors, references and the complete title (see Chapter 12).

Because of the fact that one is combining in SIMBAD the contents of various data bases, we shall call this an *integrated data base* (IDB). This name helps to distinguish it from the 'normal' data bases. The fundamental advantage of an IDB is that one replaces the secondary data of one field (e.g. the magnitudes of a radial velocity catalogue) by the primary data itself (e.g. magnitudes from a photometric catalogue); one is thus guaranteed the best available data.

It is obvious that an IDB can never be complete. It is an open structure that must accommodate new data when they arrive; cross-identifications for new objects must be provided, new parameters checked against existing ones, errors be corrected, and so on. Updating and maintenance ('plumbing') are a time-consuming and never-ending task of IDBs.

Among the problems connected with these tasks, there are two security problems. The first was mentioned in Chapter 9, namely that no user is allowed to modify the data in the DB. This is a relatively easy problem, solved through different passwords attributed to users and to 'updaters'. A more complex problem is the one of actual updating. If the data base is updated daily, the DB must be locked against users during the time the updating is done in order to prevent confusion. Such procedures are discussed in books on data bases and the reader is referred to these more technical references.

Let us return to SIMBAD. Ochsenbein (1984) quotes the following numbers for its content in 1984: 5.7×10^5 objects, of which 4.8×10^4 were non-stellar; total number of different designations 2×10^6. Similarly, for the bibliography, Carpuat *et al.* (1984) quote 1.1×10^5 objects that have at least one bibliographic reference drawn from about 4×10^5 references. This is the result of an effort extending over more than one decade.

The usefulness of an IDB becomes obvious if one considers the different possible ways of interrogating it. SIMBAD may be interrogated through:

- any designation of the object (remember there may be up to 35 different designations of a single object);
- coordinates, both equatorial (in any equinox one likes) and galactic. One may specify a box or circle around the position, within which all objects are searched and listed.

In the same way one can create samples of different types, such as a sample of stars defined by $0 < m < 6.0$, A0–A5, $\delta \geq -20°$.

Such an array of possibilities offers a large variety of applications, of which we shall consider just a few.

(a) *Monographic studies of an object*. Interrogation of the DB for a given object, for instance α Orionis, produces a list of all its designations (22), its fundamental parameters, its observational data (spectral type, radial velocity, UBV photometry, parallax, variable star characteristics and so on) and of 408 papers relating to the object (there are 125 references since 1980 alone). Only a person who has done literature searches knows what an economy in the use of time this implies!

(b) *Computer plotting of star fields*. If plotted on transparent paper at the right scale they may be superposed immediately upon photographic plates to provide identifications.

(c) *Cross-identifications*. For instance, the IRAS satellite has produced a large number of infrared sources observed at 12, 25, 60 and 100 μm. Since the position of each source is known, one can immediately check in the DB whether any coincide with a known object. Since IRAS observed over 3×10^5 point sources, cross-identification with known objects becomes a rather lengthy business and, moreover, the IRAS position and the position of the known object which lies nearby usually have different positional errors. A search of all known objects lying within, say, 20 arc sec of a given IRAS position is thus a real need, and SIMBAD can do that immediately.

(d) *Cross identifications*. The planned HIPPARCOS satellite will observe around 10^5 stars for position, proper motion and parallax. The stars selected for observation are the result of a compromise between the instrumental constraints and the wishes of the astronomers. About 200 proposals for observations of stars were made originally, including a total of 7.5×10^5 targets, but, since different proposers used different names for the same object, it was not clear how many of these targets were identical. The whole set was then fed into SIMBAD and it was found that the 5.2×10^5 targets corresponded to only 2.1×10^5 different objects. Of these, only 10^4, i.e. 5%, were not yet in SIMBAD, according to Gomez and Crifo (1985). If SIMBAD had not existed, it would have had to be created specifically for HIPPARCOS.

(e) *Field studies.* An astronomer interested in searching (say) the stars around the Pleiades for additional cluster members may request a list of all the stars in the data base, together with all known parameters. He can then concentrate his observations on the 'interesting' objects.

(f) *Field studies.* Similarly, for the spectrophotometry of galaxies, one might be interested in a list of nearby stars of known spectral type to calibrate the instrument; this again can be done immediately.

Obviously one could continue with applications, but these few samples illustrate very well the usefulness of an integrated data base, especially when compared with the old practice of consulting a (large) number of catalogues to get the same results. Because of its usefulness, such an enterprise has been envisaged for certain space projects like IRAS. To identify the sources observed by the infrared satellite, a large data base was created by entering all objects contained in some 31 different catalogues (Beichman *et al.* 1985). However, this task was done without looking for complete cross-identification between all catalogues and, as a result, two different identifiers may be listed that refer, in fact, to the same object. Although this may be satisfactory for an archive like that of IRAS, one clearly cannot enter dubious or incomplete identifications in an IDB.

It is perhaps useful to reiterate the difference between data bases like those developed for IRAS, IUE and the Space Telescope, and integrated data bases such as SIMBAD. The former are essentially the data archives of one observing facility, plus some facilities for their use. They are thus *not* intended to deal with all data on an object, nor with all kinds of objects (which often cannot even be observed from the observing facility). IDBs, on the other hand, are just the opposite: they are intended to cover all information on all objects. Whereas data archives are essentially closed, in the sense that after closing down the instrument the data base is complete, IDBs are always open. Because of their different aims, the two types of data bases should be regarded as being complementary rather than rivals.[1]

10.3.2 *Access to the information*

The existence of an integrated data base brings us back to the distribution problem. We have seen that any data base may be copied and shipped to any user. This is clearly not suitable with a continuously updated IDB because copies would need to be issued at regular intervals, say every six or twelve months. Besides the data base itself, one would also have to send the software for querying the data base, in order that it might be fully used. Such software is, however, machine-dependent and it would have to be adaptable for different computers. This complicates the problem enormously and would have been a serious obstacle to further use of the IDB; fortunately, technology found a solution through 'remote access'. This means that data stored in one computer is accessible at distant sites, perhaps several hundred kilometers away. At first this was done by means of special lines between the computer and the user but it is, of course, very expensive. The next step was to create a communication facility for the transmission of data. In analogy with a telephone or telex network, this was called a data network (DN), which may be defined broadly as a system for the transmission of data from computers to users. A data network implies both a physical and a logical structure (Albrecht 1986; Murtagh 1986). The physical structure comprises the hardware: computers, interfaces, links. This is essentially a communication facility (CF) and we shall call it this to avoid confusion.

The logical structure comprises the transmission protocols at several levels, the support software, the operational procedures and the management. Transmission protocols come in when one tries to interpret the message transmitted: although at the lowest levels this is no longer discussed, there is still no standard system for transmitting files. The support software comes in when one adapts the protocol to different machines. Operational procedures are required, for instance, to ensure that the recipient gets the message addressed to him or whether another computer can be used for this task and how this is to be done. And finally, administration is needed to define and set up the steps.

In general, communication facilities (CF) may be public or dedicated, the distinction between them being that public CFs are open to the general public and many different users, whereas dedicated CFs have specific purposes and specific users. In Western Europe, CFs are generally handled by post offices, whereas, in the United States, they are privately owned. Dedicated CFs exist, for instance, for military use, banks or multinational enterprises. Since observatories are generally economically weaker than, say, a bank, it is clear that the use of public CFs is practically the only choice open to astronomers.

Public networks being a communication facility, they work on the same economic principle as the telephone: calls are charged for, the cost depending upon time of connection and distance. The analogy with the telephone extends further in the sense that, in both cases, the users do not need to bother about technical aspects. (In the early days – and that means ten years ago – this was definitely not so.) Also, the information flow that can be transmitted is limited; the method becomes clumsy and expensive beyond certain limits. Since public CFs operate generally at about 9600 bauds (1 baud = 1 bit/s) and it is clear that any transmission that takes longer than one hour is cumbersome and expensive, there is a limit of about 2–3 Mbytes on information to be transmitted by a public CF. Out of the list of possible services, e.g.

> Message forwarding (mail)
> File transfer
> Remote log in/interactive work
> Remote job entry
> Remote processing

only a few, namely the shortest ones such as mail and short file transfer, are possible at the present time due to this restriction. To transmit larger volumes of data one needs a dedicated network, which is more expensive.

Since networks were introduced a decade ago, the situation is still in a state of flux and a number of networks are in competition, offering more or less similar services. Beyond any doubt, the next decade will see a gradual concentration: only a few will survive the rather fierce competition.[2]

In the field of astronomy, the present situation was examined at two meetings at Strasbourg, in 1985 and 1986; the transactions appeared in the *Information Bulletin of the CDS*, Nos. 30 and 31, and a further meeting is proposed for 1988.[3]

10.3.3 *Specialized communication networks*

We make at this point a short digression to mention some specialized astronomical networks, although their main concern is not so much data *per se* as data handling. The first example of such a network is 'STARLINK' (now integrated in JANET) (Lawden 1986). This British network is a set of ten similar computers located at different astronomical institutes, linked by a public CF of the packet-switching type. Its aim is to create a software environment for image and data processing. Since any existing software is shared by all users of the network, a considerable economy of manpower can be achieved in this way. The organization is such that each user group proposes the software developments needed. User groups exist for two-dimensional image processing, data base, different satellites (IRAS, IUE, EXOSAT), radio astronomy and spectroscopy. The 'software collection' is controlled by a central librarian who announces and communicates updates to the 'site managers', who are in charge of each STARLINK site. In 1985 the collection had a total volume of 5×10^5 lines of source code and occupied 10^8 bytes of disc storage.

A similar network exists in Italy (ASTRONET) (Sedmak 1986). For more details on each network, the reader is referred to the bibliography given in the Notes at the end of this chapter.

As stated before, the aim of both networks is data handling, not data as such. In consequence, neither STARLINK nor ASTRONET are data networks – they both use data collections provided by data centres. This is a very welcome division of work, which has the merit of being clear and convenient for both sides.

The subject of networks is moving very fast and, by the time this book is distributed, the situation will have improved considerably. It is easily foreseeable that, within ten years, at least all observatories of the North Atlantic community and Japan will be linked by networks that will provide each astronomer's desk (i.e. terminal) with the reduction procedures and the data and bibliographic references needed for his work. He will probably have to introduce only his own observations and a large part of the subsequent work will be done through a network. With the development of automated telescopes as we have considered in Chapter 2, we can even foresee the day when the astronomer does not even need to obtain his observations himself, but receives them on his desk. Hopefully this will liberate him for more thinking!

10.4 The future

What will be the next step in the evolution of IDBs? as we have remarked already, evolution is very rapid and any prediction will probably be wrong in a few years. Nevertheless, it seems clear that astronomers will want to have more data of all kinds.[4] There are already projects for scanning the photographic Schmidt sky survey plates with fast micro-photometers (McGillivray *et al.* 1988) so that a numerical image of the plate is produced that can be stored on optical discs. Such a survey, down to 19th magnitude, will probably contain information on some 10^8–10^9 objects (stars and extended objects) and will provide us with the sky image at visible wavelengths. Similar surveys exist already at other wavelengths, for instance at radio wavelengths. They contain a smaller number of objects because in the radio band, resolution is coarser than in the optical region by a factor of 10^2 or more.

We can see the direction in which work will probably go: the collection and distribution of surveys made at several wavelengths. The next step is then to produce object images at different wavelengths, superposable on a screen – for instance infrared clouds projected upon star maps or upon radiomaps – to find out what relations exist between them.

Due to the volume of information to be handled, it is not yet clear whether existing data centres can store the information and distribute it to users over public networks, or whether each observatory must buy its own set of 'sky surveys' (on optical disc). Since this depends very much on a rapidly evolving technology, it seems out of the question to venture predictions.

Notes on Chapter 10

The subject of this chapter has been treated in two IAU Colloquia, namely No. 35, *Compilation, critical evaluation and distribution of stellar data* (Jaschek and Wilkins 1977) and No. 64, *Automated data retrieval in astronomy* (Jaschek and Heintz 1982).

On the same subject there is also the proceedings of a meeting at Tbilisi (Kharadze 1985) and those of an international course on *Data handling in astronomy and astrophysics* (Hauck and Sedmak 1985), both of which contain a description of the later developments. The latest meeting on the subject was *Astronomy from large data bases*, in October 1987 (Murtagh and Heck 1988).

1. The difference between data centres and archives is analysed in detail in Jaschek (1988). A list of computer-readable space astronomy archives is given in White (1987).

2. Just to mention at what considerable speed the use of networks is proceeding, let us quote the number of remote users of the CDS database. We had 20 users in 1984, 42 in 1985, 69 in 1986, 82 in 1987 and are at 113 in July 1987. The CDS is being used by astronomers in Hawaii, South Africa, Australia and Haifa; in fact there are twice as many users outside France as inside.

3. On data networks in astronomy, the best summary paper is Murtagh (1986).

For some recent trends in what non-stellar astronomers wish to get in the future from data centres, see also Egret and Guibert (1984). Legal aspects are discussed in Jaschek and Heintz (1982).

References

Albrecht, R. (1986) *BICDS* **30**, 35

Beichman, C. A., Neugebauer, G., Habing, H. J., Clegg, P. E. and Chester, T. J. (ed.) *IRAS Explanatory Supplement*, Joint IRAS Science Working Group

Carpuat, C., Damge, M., Dubois, P., Kirchner, S., Lagorce, A., Laloe, S., Ochsenbein, F. and Wagner, M. J. (1984) *BICDS* **26**, 83

Dubois, P., Ochsenbein, F. and Paturel, G. (1983) *BICDS* **24**, 125

Egret, D. (1983) *BICDS* **24**, 109

Egret, D. and Guibert, J. (1984) *L'avenir des données non-stellaires*, Observatoire de Strasbourg

Gomez, A. and Crifo, F. (1985) In *Hipparcos: Scientific aspects of the input catalogue preparation*, ESA SP-234, p. 57

Hauck, B. and Sedmak, G. (ed.) (1985) *Data handling in astronomy and astrophysics, Mem. Soc. Astron. Ital.* **56**, No. 2 and 3

Jaschek, C. (1988) In *Astronomy from large databases* (ESO Conf. and Workshop Proc. **28**), Murtagh, F. and Heck, A. (ed.), ESO, Garching

Jaschek, C. and Wilkins, A. G. (ed.) (1977) *IAU Coll. 35, Compilation, critical evaluation and distribution of stellar data*, Reidel

Jaschek, C. and Heintz, W. (ed.) (1982) *IAU Coll. 64, Automated data retrieval in astronomy*, Reidel

Jung, J. (1971) *BICDS* **1**, 3

Kharadze, E. K. (ed.) (1985) *Stellar catalogues: data compilation, analysis, scientific results*, Abastumani Bull. **59**

Lawden, M. D. (1986) *BICDS* **30**, 13

McGillivray, H. T., Dodd, R. J. and Beard, S. M. (1988) In *Astronomy from large databases* (ESO Conf. and Workshop Proc. **28**), Murtagh, F. and Heck, A. (ed.), ESO, Garching

Murtagh, F. (1986) *BICDS* **31**, 89

Murtagh, F. and Heck, A. (ed.) (1988) *Astronomy from large databases* (ESO Conf. and Workshop Proc. **28**), ESO, Garching

Ochsenbein, F., Egret, D. and Bischoff, M. (1977) In *IAU Coll. 35, Compilation, critical evaluation and distribution of stellar data*, Jaschek, C. and Wilkins, A. G. (ed.), Reidel, p. 31

Ochsenbein, F., Bischoff, M. and Egret, D. (1981) *AA Suppl.* **43**, 259

Ochsenbein, F. and Bischoff, M. (1982) In *IAU Coll. 64, Automated data retrieval in astronomy*, Jaschek, C. and Heintz, W. (ed.), Reidel, p. 211
Ochsenbein, F. (1984) *BICDS* **26**, 75
Sedmak, G. (1986) *BICDS* **30**, 13
White, N. E. (1987) In Appendix C of *Astrophysics Data Program* NRA–87–OSSA–11

11

The publication of scientific information

In this chapter we shall deal with the ways scientific information is published. Broadly speaking scientific information can be divided into three types, namely: theories, techniques and data: the first provides the framework, the second shows how to confront theories with observations and the third summarizes the results. It is clear that scientific information if not communicated is useless for the development of science, because in one sense it is 'non-existent'. Take, for instance, the technical inventions of Leonardo da Vinci (1452–1519): since they were not published in his time, but only in the twentieth century, they had no impact upon the development of technology. Proper publication of research results is thus a prime concern for any scientist because, in this way, information becomes knowledge that can be criticized and verified by other scientists and may become an accepted part of science.

11.1   Historical development

11.1.1 *From books to journals*

Greek and Arab astronomers published their results in the form of handwritten *books*, written on papyrus sheets or parchment. Since usually only a few copies of each book existed – each copy implying the work for weeks of a copyist – it is no wonder that few books have survived. Scholars estimate that only 10 or 20% of all existing books of classical antiquity have survived and this is also true for astronomy. The biggest chance of survival lay with the books that were addressed to a wide audience, like Aratos' *Phaenomena*, a poem that describes the constellations, or a compendium like the *Syntaxis* (or *Almagest*) of Claudius Ptolemaeus; most highly technical works did not survive. A further problem for survival was the relative fragility of papyrus sheets in humid climates; books therefore had to be recopied to stay readable and, in this way, a number of errors slipped in.

Some of the difficulties were overcome with the introduction of paper (invented by the Chinese) by the Arabs during the fourteenth century. Paper proved to be quite durable and cheap. The next important step was the invention of printing by Gutenberg (Strasbourg and Mainz) in the fifteenth century, which essentially introduced the book in the form we know it today. The advent of the printed book was a true revolution for scientific communication because it offered the possibility of rapid production of many copies on a stable and durable support medium. At the beginning, printed books were however expensive; due to the lack of a large number of scientists capable of buying them, it is no wonder that scientific books were slow to appear. Among the books published in the fifteen century, only 10% or so are books on science. A perusal of the titles published before 1500 (Grassi 1977) shows that the most widely printed book on astronomy was the *Sphaera mundi* by Sacro Bosco (John Hollywood). Yearly almanacs occupied a large place, as well as translations of classical Greek and Arab authors. But, in all, there were only about 100 books published on astronomical (and astrological) matters in about 30 years. This can be compared (see Chapter 12) with the 300 or so books on astronomy published each year now.

All important scientific results were published in the following centuries in the form of *books* – be it the works of Tycho Brahe, Kepler, Newton or Gauss. However, in the nineteenth century, the book gradually lost its place as the carrier of new theories in physics and in astronomy. A book like *The origin of species*, by Darwin, which announced a new theory, is difficult to conceive today, except in social sciences and humanities. The origin of this gradual displacement of the book lies in the fact that a book is long (over a hundred pages generally); therefore, it takes time to be written and, during that time, other scientists may anticipate one's discoveries. On top of that, the technical production of the book i.e. the transition from the manuscript to the printed book, takes additional time. In an age of ever-increasing publication speed, the combination of these two factors became lethal for the role of the book as disseminator of discoveries.

The main function of the book nowadays is to present a unified view of past discoveries. This change can best be illustrated by the time lag between the discovery of a new type of object or phenomenon and when the first book on the subject was published. A glance at Table 11.1 shows

Table 11.1. *Time delay between the discovery of a phenomenon and the first book on the subject*

Phenomenon	Discovery by	date	First book by	date
Degenerate stars	Adams	1914	Schatzman	1958
Interstellar matter	Trumpler	1930	Spitzer	1968
Flare stars	Luyten	1949	Gurzadyan	1980
Pulsars	Hewish *et al.*	1968	Manchester and Taylor	1977

the time lag may be quite considerable. A good book on a subject is, on the other hand, very often a landmark for further developments of the subject and as such is irreplaceable. But let us return to the historical developments.

Besides books, shorter notices were also published, describing, for instance, the apparition of a comet or a nova, the occurrence of meteors, solar eclipses, aurorae – in short all the phenomena that Aristotle called 'meteorological', i.e. belonging to the sublunar world of transient phenomena. Such pamphlets normally consisted of a few printed pages, frequently containing a figure; sometimes they reached the size of a booklet. Since no good distribution networks for such items existed, the distribution of pamphlets was difficult and each scientist maintained a steady correspondence with his colleagues to keep them informed. The correspondence between scientists was an important channel of communication from the end of the fifteenth century onwards. Its importance decreases only when more regular channels for communications were established. (Today its role has been taken oven by the telephone.) In the seventeenth century, the number of scientists had grown so much that the exchange of letters became impracticable. The need for more efficient communication was satisfied by a new type of publication – the proceedings of learned societies. In France, the *Journal des Scavants* – the first scientific journal – was started in 1665; it had a short life and was superseded by the *Proceedings of the Academy of Sciences* (from 1666 on). The *Philosophical Transactions* of the Royal Society of London started the same year. These two learned societies were joined in the next decades by a growing number of Academies of Sciences of other countries. Proceedings were published at relatively short intervals (usually less than a year) and exchanged with those of other Academies. Thus a more efficient exchange of communication started, which generated a rapid growth in the volume of information.

At the beginning, results from all sciences were published together but, very soon, a system of specialization set in, which resulted in a division of the 'proceedings' publications into various series, each one relating to a special group of sciences.

11.1.2 *Peer review*

The most important feature of this new system of publication was the procedure for accepting a submitted paper. For the *Philosophical Transactions* it was decided that one of the secretaries (H. Oldenburg) was to be in charge of publications and that a paper could be accepted for publication after revision by one of the members of the Council. The French Academy followed a similar system, with the only difference that any academician could act as referee. Such practices introduced the 'peer review system', which filters submitted papers through the advice of capable specialists. The system guarantees that the content of the paper satisfies scientific standards, a guarantee that did not exist for books.

The peer review system was later adopted by all journals and is still in use today, a fact which speaks in favour of its usefulness. (The reader can find a detailed account of it in Merton (1973).) Nowadays, the 'opinion of one of the members of the Council' is replaced by that of a referee, usually anonymous, chosen by the editor. The anonymity is a precaution to avoid polemics between the author and referee, who occasionally differ sharply in opinion on the value of the paper!

The referee may accept the paper as it is presented, suggest modifications or reject it. In the second case, a modification of the paper usually leads to its acceptance. If the paper is rejected, the author can either ask for a second referee if he wants to publish in the same journal, or send the manuscript to another journal. Although rejection is frequent, it is not always a sign that the paper is bad – it can also mean that it 'falls outside current ideas', whatever that might mean in a specific case.

The percentage of manuscripts rejected depends very much on the subject area – in humanities it is close to 80%, in mathematics about 50% and in physics about 25% (Merton 1973). In one astronomical journal, *Astronomy and Astrophysics*, the percentage is 23% (Lequeux 1980), which is in line with physics. It should, however, be added that many of the rejected papers appear later in a modified form somewhere else, so that these figures should not be interpreted as implying that 23% of the astronomical papers written are barred from publication.

11.1.3 *Growth of periodicals*

Let us return to the course of events. As we said earlier the *Philosophical Transactions* were followed by a number of other scientific periodicals, so that the total number grew rapidly. Fig. 11.1, taken from de Solla Price (1963), shows the trend. After a somewhat irregular start, the number grew exponentially over at least two centuries. (We shall consider the quantitative aspects of growth in detail in Chapter 12.)

In the late eighteenth century, one finds a new phenomenon, namely, the privately run scientific journal. Their appearance was due to an ever-increasing number of scientists capable of feeding and economically sustaining a journal; these journals are necessarily restricted to one science or part of a science. Because of this, it is clear that specialized journals started at different epochs for different sciences; the important factor was the number of scientists in each speciality. In astronomy, the oldest

Fig. 11.1. Total number of scientific journals and of abstracting journals, as a function of time, from de Solla Price (1963).

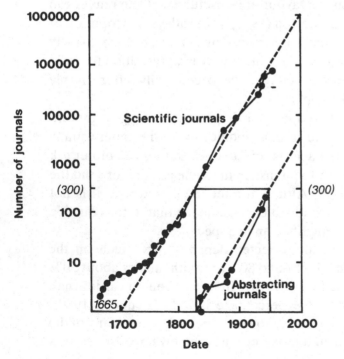

surviving journals are the *Astronomische Nachrichten*, founded in 1823, which is the offspring of a series of publications coming from the eighteenth century, and the *Monthly Notices of the Royal Astronomical Society*, founded in 1831. Fig. 11.2 shows the trend in astronomical journals; as we can see, it also follows an exponential like that of Fig. 11.1. This figure should be also compared with Fig. 2.1, which gave the number of observatories as a function of time. It should be added that these private journals were (and are) often the organ of a professional scientific society; these societies gradually replaced the academies of the preceding century.

11.1.4 *Bibliographies and abstracts*

With the multiplication of journals it became more and more difficult for each scientist to know all relevant papers published and therefore the logical next step were 'bibliographies', in which all relevant papers of a discipline are listed. Although the first bibliographies of this type belong to the eighteenth century, they really started to multiply in the nineteenth century. The reason is that by 1830 about 300 scientific journals existed and the bibliographies became essential for keeping up-to-date. These bibliographies can also be called 'abstracting journals' because, besides the purely bibliographic references, a short abstract of the content of the publication is also given.

The history of abstracting journals runs strictly parallel of that of journals (see also Fig. 11.1): there is exponential growth here too.

Fig. 11.2. Number of journals reporting astronomical research. Data for 1825–75 from Houzeau and Lancaster (1964), for 1900–75 see Table 12.1. The line represents: $N = 34 \exp [0.0123\,(t - 1825)]$.

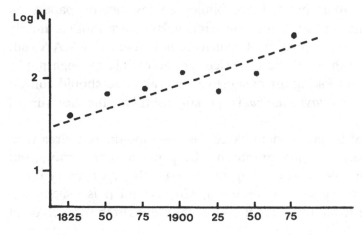

In astronomy there are three major bibliographies (Schmadel 1979):

   (a) J. Lalande, *Bibliographie astronomique* (1803), which covers the period before 1800;

   (b) J. C. Houzeau and A. Lancaster, *Bibliographie générale de l'astronomie*, Bruxelles 1887–1892. New edition, London 1964. Covers the period before 1881;

   (c) *Astronomischer Jahresbericht* (AJ), Berlin 1899–1948, Heidelberg 1948–1969, which was succeeded in 1969 by *Astronomy and Astrophysics Abstracts* (AAA), Heidelberg. Both AJ and AAA are published by the Astronomisches Rechen-Institut.

The AAA is nowadays produced in two volumes per year, thus reducing the delay between the publication of a paper and its inclusion in AAA to less than one year. Most abstracts are given in English and, except for abbreviation in the case of very long abstracts, AAA reproduces preferentially the author's own abstract. All papers listed are classified according to subject, using a rather well-adapted classification system taken over from the AJ and perfected over the years. There is also an author index.

AAA is now also available in computer-readable form, but only from 1985 on. This is probably the greatest drawback of the present AAA.

Besides AAA, some national abstracting services also cover astronomy, namely INSPEC in the United Kingdom, the *Bulletin Signaletique du CNRS* in France, FIZ (*Fach-Informations-Zentrum für Physik*) in the German Federal Republic, and the *Referativnyi Zhurnal* in the USSR. Since these services cover many areas other than astronomy, they do not include as many astronomical journals as AAA and are thus less useful than AAA if one is looking for complete coverage of astronomical subjects. On the other hand, these bibliographies include papers on astronomy published in non-astronomical journals (such as those primarily on physics, chemistry, history, etc.), which are not included in AAA and, moreover, the first three of these services are accessible through public networks. If one is looking for exhaustive coverage one should consult them too. The *Referativnyi Zhurnal* is specially useful for the literature of Eastern countries.

Another type of bibliographic service has been mentioned already in Chapter 10, namely the bibliography not by paper and/or authors, but ordered by object. As we have seen, many research papers carry information referring to various astronomical objects and it is usually impossible to deduce from the title of the paper the names of the objects dealt with.

A French group headed by R. Cayrel started in the nineteen-sixties to compose a bibliography by object covering the literature from 1950 on. At first, the only astronomical objects considered were relatively bright stars and only a small number of journals was analysed. This has changed over the years. (For a detailed history of the bibliography see Ochsenbein (1982) and Dubois and Ochsenbein (1983).) The service now includes all kinds of astronomical objects and covers a large number of journals. The work is carried out as a collaboration between the Institut d'Astrophysique at Paris, the Bordeaux Observatory and the CDS. Up to now it contains more than 6×10^5 references relating to over 1.5×10^5 objects, with an average of four citations per object. For each paper, the complete bibliographic reference, including the explicit title, is provided. This bibliography is updated continuously and is accessible on line, or on microfiche. The microfiches cover periods of whole years, but their publication is out of date, in the sense that in 1986 only the microfiches up to 1983 were available.

Another bibliography that is very useful is the *Science Citation Index* (SCI). It covers 2500 journals worldwide, including 40 astronomy titles. It is strictly a bibliographic service and carries no abstracts. It is divided into three series, namely the Source Index, the Citation Index and the Permuterm Subject Index. The first part is an index of current journal papers, with author, title, references and volume/page/date information. The Citation Index is an index of articles cited by other authors. We shall refer in Chapter 12 to some of the uses of the SCI for the evaluation of scientists.

For completeness, I should also mention the bibliographies compiled on very specific subjects, such as that by Davoust and Pence (1983) on surface photometry of galaxies, or the bibliographic catalogue on variable stars by Huth and Wenzel (1985, 1986). Such bibliographies can be extremely useful in a particular field. I list in Table 11.2 some of the bibliographies available on tape from the data centres. (It would be a useful exercise to publish a list of all existing bibliographies, either printed or computer-readable.)

11.1.5 *Attempts to reduce publication delay*
After this excursion into bibliography let us now go back to the general development of astronomical publications. We have seen that one of the inconveniences of books is the long time delay between their inception and publication; journals were invented to speed this up. Sample studies carried out for two journals (*Astrophysical Journal* and *Astronomy*

Table 11.2. *Machine readable bibliographies on specialized subjects*

Bibliographic catalogue of field RR Lyr stars (Heck and Lakaye 1977)
Bibliographic index of planetary nebulae (Acker, Marcout and Ochsenbein 1981)
Bidelman's stellar data file (Parsons, Buta and Bidelman 1980)
Bibliographic catalogue of variable stars (Huth and Wenzel 1985) and *Supplement*
    (Huth and Wenzel 1986)
Annotated bibliography of multivariate statistical methods in astronomy (Murtagh
    and Heck 1986)
Catalogue of star clusters and associations (Ruprecht, Balasz and White 1982
    (*Associations*) and 1984 (*Globular clusters*))
Detailed bibliography of surface photometry of galaxies (Davoust and Pence 1982)
A catalogue and bibliography of Mn–Hg stars (Schneider 1981)

---

and Astrophysics) show that the average delay in 1985 was ten months between the reception and the publication of a paper.

Since such a delay is now often considered to be too long, new series of publications have appeared, either within journals ('Letters to the Editor') or in the form of special journals like *Astrophysical Letters*. Their main purpose is to diminish the publication delay; typically, it is reduced to a few (3–6) months. In the case of *Nature*, a British publication covering all sciences, the delay is only of the order of a few weeks; the same is true for the notes published in the *Comptes Rendus de l'Academie de Sciences* (Paris).

11.1.6 *Observatory publications*

After dealing with books and journals, which are the best known ways of publication, let us now consider a third type, namely, 'observatory publications'.

An 'observatory publication' is a publication of an observatory that presents the research results of its staff members. This type started with the birth of the first modern observatories at the end of the seventeenth century and was considered for at least two centuries to constitute the 'proper way' to publish the observations of staff astronomers.

The system fell slowly out of favour because observatory publications are non-periodic and the time elapsing between submission and publication is usually much longer than in journals. Furthermore, they turn out to be more expensive since printing and distribution have to be done by the institution concerned, whereas a journal provides a specific organization to do this. Observatory publications are now reduced to the publication of

Table 11.3. *Epoch at which some observatories suspended their publication series*

Observatory	Date	Observatory	Date
Yerkes	1960	Lick (Contrib.)	1973
Leander McCormick	1960	Paris	1970
Michigan	1955	Kitt Peak	1970
Haute Provence	1970		

very long series of observations or data, which are difficult to publish in journals, such as astrometric observations. Other exceptions are the observatories in Socialist countries, which are generally still keeping their series alive. To show how the situation has evolved, let us consider the data assembled in Table 11.3, which provide the dates at which some observatories stopped their 'observatory publications'. It is to be noticed that some of these series were replaced by publications of offprints from journals, but the introduction of photocopiers (Xerox type) in the nineteen-fifties made such series superfluous. The only positive aspect of offprints is that they can provide easy access to papers published in journals that are not in one's own library.

The decline or disappearance of observatory publications has, however, been more than compensated for by the appearance of some new types of publications – the proceedings of meetings and minor publications such as newsletters and internal reports.

11.1.7 *Proceedings of meetings*

Let us begin with astronomical meetings. The first international astronomical meeting was organized in Germany in 1798 by v. Zach and immediately ran into difficulties with the authorities, since a French delegate (Lalande) was to be present. France had just experienced the Revolution of 1789, and Frenchmen were thought to be highly subversive! The meeting was nevertheless held at Gotha and was attended by fifteen astronomers; it was the first international meeting. During the nineteenth century such meetings were rare, mostly because travel was slow and difficult. But when large cooperative efforts started, like the 'Astrographic Catalogue' organized by French astronomers, meetings become a necessity. The number of meetings began to increase rapidly after the 1890s and became very large after World War II, in the wake of the general facilities for air travel. Besides such international meetings, a number of astronomical societies hold regular national meetings, which are frequently also attended by astronomers from abroad.

The increase in meetings is of course not a specifically astronomical phenomenon; as usual, astronomy only participates in the general trend. For instance, in 1975 there were 2846 international meetings, organized by international bodies, a number which in 1978 had grown to 3727. Of this number, scientific meetings constitute only about 10%, and ones on astronomy a small fraction of this. In 1986, about 367 international scientific meetings were held cosponsored by the International Council of Scientific Unions (ICSU) (see Chapter 13).

The trend to hold more meetings must be considered as an expression of the need for increasingly rapid communication between relatively small groups of specialists within an ever-increasing number of astronomers. The 'latest news' can thus be communicated with minimum delay.

Over the years, a loose nomenclature has developed for meetings: if the meeting is attended by more than 150 people, it is called a 'symposium'. A 'colloquium' has 50–150 participants and workshops are usually meetings of about twenty persons. The word 'conference' does not imply any specific number.

Usually, each meeting produces a volume of proceedings. Since it is desirable to get it out quickly, the printing is usually done by reproducing a typed manuscript directly; such a procedure means that the proceedings can be published within one year of the meeting. Papers communicated at meetings are, in general, not subject to prior peer review and, unavoidably, some lower-quality papers find their way into the proceedings. Because of this and because of the emphasis put on rapid communication (which produces ephemeral papers), communications given at meetings are rightly considered to have less weight than papers published in journals.

A good feature of meetings (and of the proceedings) are the so-called 'review papers' or 'invited papers', which originate from an invitation made by the organizer(s) of the meeting to a well-known specialist to summarize or review a narrow subject of the meeting. These invited papers are usually the best feature of the proceedings.

11.1.8 *'Grey' literature*

Another type of publication is the so-called Newsletter, which is edited for a rather small audience of specialists (typically about 200–300), provides the latest news and is often of ephemeral value.

Internal reports are documents produced within big organizations (e.g. NASA, ESA, ESO), also for a restricted audience. Since they are often concerned with the very latest matters they are circulated in restricted numbers to outsiders. Because they are not for general distribution, they go only to the small circle of specialists of the subject. They should not be quoted as references, if possible.

Librarians use the term 'grey literature' for what I call 'minor publications' and include under this term reports, technical notes and specifications, preprints, translations, official publications (by governmental bodies), supplementary publications, trade literature, dissertations and theses. Chillag (1985) characterizes 'grey literature' as items with poor availability, poor bibliographic information and control, non-professional layout and format and low print runs. The interested reader is referred to this author for more details.

The term 'grey literature' is used in contrast with 'primary literature', which comprises books and periodicals.

11.2   Forms of publication

As we have seen, the traditional material for publications has been paper (see Chapter 4). Paper is cheap and it is durable. If the paper (and ink) quality was good and the conditions of storage reasonable, printed papers are easily preserved for three or four centuries.

One inconvenience is that printed papers are bulky, because the size of type cannot be reduced beyond certain limits. The unit for type size is the 'point' (= 0.31 mm). Ten-point type (in fact anything in the range 8–12) is considered a normal size, whereas 5-point type is difficult to read. The consequence is that printed publications occupy ever-increasing numbers of shelves in any library. A second inconvenience of printing is that the composition of typescript is lengthy and therefore expensive. Solutions to both difficulties are actively sought.

An easy solution is the typing of manuscripts with modern text-processing facilities. This avoids manual composition and decreases the cost. Such a procedure is often adopted for proceedings of meetings; it shifts the burden of composition from the printer to the author. Since each author usually has a different processor, the books produced this way are not very aesthetically pleasing. The process also avoids proofreading but, equally, it prevents any intervention of the editor in the manuscripts. For books and journals this is therefore not a usable solution.

To overcome the bulk problem, one immediate solution is to diminish the size of the print, from 10 to 1–4 points. This is called microprint. More information goes on a printed page, but readers must have excellent eye-sight or a magnifying glass. Moreover, the gain is not very great.

The next step is the microfiche, which is a plastic transparent film sheet upon which the text is reproduced. Each microfiche (148 × 105 mm), can contain up to 220 pages of text; it is also cheaper to produce than a printed text. The inconvenience is that to read a microfiche, one needs a magnifying device (microfiche reader) that enlarges about 35 times. In general, one cannot say that microfiches have had the overwhelming favour of the public, as far as journals are concerned. On the other hand, for data, microfiches are a cheap solution – one can carry on microfiches weighing 30 g the equivalent of seven printed volumes weighing over 10 kg.

A further step can be that of putting the whole text in machine-readable form, for instance on magnetic tape. In doing so one can compress the information further but the printed listing of a tape is often more bulky than regularly printed pages. Furthermore, one needs a computer to convert from the tape to a human readable text.

A probable next step is the 'electronic journal', in which papers are no longer printed but appear on a screen. If a paper is considered to be interesting, a printout can be obtained. Although this may be a good solution for the reader, it makes the literature stored in this way not permanent: as we have seen, tapes have a lifetime of less than ten years. Unless one gets a better material support, electronic journals seem premature, at present time.

Up to now we have dealt with scientific publications in general; we shall next consider the publication of data in particular.

11.3   The publication of data

From what has already been said, one can surmise that data are published primarily in journals, also in observatory publications and occasionally in books. Journals constitute the primary form of data publication. Most frequently this is done in the form of short lists of data; for stellar photometry, for example, it has been found that, in the 1980s, about two hundred articles per year were published with less than twenty objects per paper. Longer lists (more than a hundred items) are called catalogues and usually appear as observatory publications or, if published in journals, in the form of an announcement that a new catalogue has been deposited at a data centre (see Chapter 7), from which it can be obtained.

Books usually do not carry data and, if they do so, the data are not the latest ones. Exceptions are critical data compilation works, where standard values or standard relations are given. Such data compilations are basic for all fields, including astronomy. The most useful ones available are:

- C. W. Allen, *Astrophysical quantities* (*third edition*), Athlone Press, London, 1974; the book is now somewhat outdated and a new edition is urgently needed.
- The volumes corresponding to astronomy of the Landolt-Börnstein series *Zahlenwerte und Funktionen*, Berlin, 1982. These volumes are more up to date, but suffer from a rather complicated system of ordering.

By the way, the history of the Landolt-Börnstein series reflects the data explosion well. The collection was started in 1883, as a one-volume book with 261 pages. The latest edition (1960–86) contains 98 volumes with a total of 52 000 pages!

A series of parameters of standard stars is given in a forthcoming publication by A. G. Davis Philip and D. Egret in the series *Publication Speciale du CDS*.

For reference data in physics and chemistry, the recommended book is the *CRC Handbook of chemistry and physics* published by Chemical Rubber, which is updated annually. If more specialized tables are sought, one should use the *Handbooks and tables in science and technology, second edition*, edited by R. H. Powell, Oryx, 1983.

Astronomers usually complain about the large data flow in astronomy, which makes life difficult. When compared to chemistry and physics, the situation in astronomy is, however, not alarming. In chemistry and physics, the flood of data is so large that three new types (new for astronomers) of publication have had to be introduced, which I shall mention briefly.

First, we have the data journals. These are journals dedicated exclusively to data in a given field. They accept data from individual authors consisting primarily of their own measurements. As examples we may quote the *Journal of chemical and engineering data* and the *Atomic data and nuclear data tables*. Although nothing similar exists as yet in astronomy, such a journal will certainly be needed in the next few years.

Then we have the reference data journals, which provide compilations of data from a comprehensive range of primary journal articles. The author of the compilation exercises a critical evaluation to derive the 'best' data. As an example, we may quote the *Journal of physical and chemical reference data*. Such a journal can be seen as a continuous update of the critical data compilations mentioned above.

Finally, there are the synoptics, which carry only summaries of one or two pages; the full papers are published in other journals in miniprint or microfiche, or left in a depository from where they can be extracted at moderate cost.

The only astronomical attempt to implement one of these new types of information storage is the IAU Commission 27 (Variable Stars) file of unpublished photoelectric observations of variable stars (Fitch 1971). The archives are kept at the following three places:

(a) The Royal Astronomical Society, Burlington House, Piccadilly, London, W1V 0NL, UK.
(b) Odessa Astronomical Observatory, Shevchenko Park, Odessa 270014, USSR.
(c) Centre de Données Stellaires, 11, rue de l'Universite, 67000 Strasbourg, France.

Copies of any file can be obtained from any one of them. There are now about 200 files. The coordinator of this archive is M. Breger, Universitäts-Sternwarte, Türkenschanzstr. 17, A–1180 Wien, Austria. These archives are not consulted very often.

The fact that, despite everything that is said about the information flood in astronomy, we still have no data journals or data reference journals shows that, after all, we are still handling a reasonable amount of data.

11.4   How to search for relevant literature

We shall devote the last section to the practical problem often encountered by students of astronomy: how to search for the appropriate literature in an unknown field. (Data are omitted because of the discussion already given in Chapter 5 and in this chapter.)

The first step is to consult a text book. Such books usually provide few technical details, but place the subject in a wider context. A good guide for finding an appropriate text-book is *A bibliography of astronomy 1970–79* by R. A. Seal and S. S. Martin. This book also provides a list of monographs on more specialized fields, which is the second step in the literature search. Monographs provide more details than textbooks but suffer from the drawback of all books, a gap of at least two years between inception and publication, so that monographs do not contain the very latest developments in the field.

The third step is to look for a recent update, which can be either a summary paper (also called review paper) given at a recent meeting or a summary paper of the type published in the series *Annual review of astronomy and astrophysics* (ARAA) published by Annual Reviews Inc., Palo Alto. This is an American series in which one volume per year is published containing about twenty summaries of different topics. These summaries are very up-to-date and provide an extended list of bibliographic references. Being an American publication, it has a natural bias toward authors living in the United States and work done there.

In Europe, there are three series that parallel the *Annual Review*, namely, *Vistas in astronomy*, published by Pergamon Press, and the series of summary papers in *Mitteilungen der Astronomischen Gesellschaft*. Both of these series are, however, more limited in scope than ARAA. *Vistas in astronomy* also carries many historical contributions besides the summary papers. The third series is *Space science review*, published by Reidel. To locate review papers, readers should use the *Astronomy and astrophysics abstracts* (AAA), which, as I have said already, is ordered by subject and is a means of locating rapidly the interesting items. One should start with the latest issue of AAA and work backwards in time.

Only after general books, monographs and review papers on the subject have been consulted, is the reader ready to read current research papers. These can be found with the help of the AAA, or by browsing through recent issues of journals.

Finally, to make sure that one has not overlooked significant research in the area, one can use the *Reports on astronomy* published by the IAU (see Chapter 13) every three years; all fields of astronomy are covered in a very comprehensive way.

The procedure of starting with the general and ending up at the particular, i.e. the research papers, is definitely superior to approaching the problem the other way around, namely by starting with reading research papers. At first one usually lacks the critical judgement needed to deal with the sometimes conflicting evidence presented by the authors in their papers. Furthermore, at a time when the trend of editors is to condense the papers as much as possible, research papers presuppose good prior knowledge of the subject. Most research papers today are unintelligible to outsiders of the field.

I have left 'grey literature' to the end. This literature is generally not useful for the beginner. It becomes vital for the specialist, who finds in it his main source of the very latest news; he becomes an avid consumer of preprints, newsletters and so on. Such material is usually published six to ten months later as a paper in a regular journal; by reading the preprints, the user gains that much time over colleagues who only read journals. For frontier research this may be crucial.

Notes on Chapter 11

The outline of the history of publications in astronomy given in this chapter is a summary of scattered notes found in books on the history of science and in particular in those on astronomy. A complete history remains to be written, however.

A good general introduction for general scientific literature is de Solla Price (1963). An excellent survey of the information in physics is given in Shaw (1985). This book covers very well the general matters that are common to astronomy and physics and constitutes a very useful reference book. Astronomy is dealt with partially in a chapter covering geophysics, astrophysics and meteorological physics, by E. Marsh.

A book that provides essential background on ancient astronomical publications is Neugebauer (1975). For modern times we have nothing equivalent. A study of the crucial period around 1800, which saw the appearance of the first astronomical journals is Hermann (1972). An excellent summary of commercially-available bibliographic services was given by Rey-Watson (1983). An update of this paper is in the press (1988).

References

Acker, A., Marcout, J. and Ochsenbein, F. (1981) *AA Suppl.* **43**, 265
Adams, W. S. (1914) *PASP* **26**, 198
Allen, C. W. (1974) *Astrophysical quantities, third edition*, The Athlone Press
Chillag, J. (1985) In *Information sources in physics, second edition*, Shaw, D. F. (ed.), Butterworth, p. 355
Davis Philip, A. G. and Egret, K. (1988) CDS, to be published
Davoust, E. and Pence, W. (1982) *BICDS* **25**, 9
Dubois, P. and Ochsenbein, F. (1983) *BICDS* **27**, 7
Fitch, W. S. (1971) *IAU Information Bulletin on variable stars* No. 510
Grassi, G. (1977) *Union catalogue of printed books of the XV and XVI centuries in European astronomical observatories*, Rome Astr. Obs. Library
Gurzadyan, G. A. (1980) *Flare stars*, Pergamon Press
Heck, A. and Lakaye, J. M. (1977) *AA Suppl.* **30**, 397
Hermann, D. B. (1972) *Die Entstehung der astronomischen Fachzeitschriften in Deutschland, Veröff*. Archenhold Sternwarte No. 5
Hewish, A., Bell, J., Pilkington, D. H., Scott, P. F. and Collins, R. A. (1968) *Nature* **217**, 709
Houzeau, J. C. and Lancaster, A. (1964) *Bibliographie générale de l'Astronomie*, The Holland Press
Huth, H. and Wenzel, W. (1985) CDS catalogue 6035
Huth, H. and Wenzel, W. (1986) CDS catalogue 6038
Lalande, J. (1803) *Bibliographie astronomique*, Paris

Lequeux, J. (1980) *Journal des Astr. Français*, No. 9, p. 1
Luyten, W. (1949) *Ap. J.* **109**, 532
Manchester, R. N. and Taylor, J. H. (1977) *Pulsars*, Freeman
Merton, R. K. (1973) *Sociology of science*, University of Chicago Press
Murtagh, F. and Heck, A. (1986) *AA Suppl.* **68**, 113
Neugebauer, O. (1975) *A history of ancient mathematical astronomy*, Springer
Ochsenbein, F. (1982) In *IAU Coll. 64, Automated data retrieval in astronomy*,
    Jaschek, C. and Heintz, W. (ed.), Reidel, p. 171
Parsons, S. B., Buta, N. S. and Bidelman, W. P. (1980) *BICDS* **18**, 86
Rey-Watson, J. M. (1983) *Information sources and services in astronomy,
    astrophysics and related space sciences*, Smithsonian Inst. Libraries Research
    Guide No. 2
Rey-Watson, J. M. (1988) In *Astronomy from large databases* (ESO Conf. and
    Workshop Proc. **28**), Murtagh, F. and Heck, A. (ed.), ESO, Garching
Ruprecht, J., Balasz, B. and White, R. E. (1982) *BICDS* **22**, 132
Ruprecht, J., Balasz, B. and White, R. E. (1984) *Soviet Astronomy* **27**, 328
Schatzman, E. (1958) *White dwarfs*, North Holland
Schmadel, L. D. (1979) *BICDS* **17**, 2
Schneider, H. (1981) *BICDS* **20**, 113
Science Citation Index, Garfield, E. (ed.), Inst. for Scientific Information
Seal, R. A. and Martin, S. S. (1982) *A bibliography of astronomy 1970–79*,
    Libraries Unlimited Corp.
Shaw, D. F. (ed.) (1985) *Information sources in physics, second edition*,
    Butterworth
Solla Price, D. de (1963) *Little science, big science*, Columbia University Press
Spitzer, L. (1968) *Diffuse matter in space*, Wiley
Trumpler, R. J. (1930) *PASP* **42**, 214

12

The growth of scientific information

In the previous chapter we examined the ways in which information is published. We shall examine in this chapter some quantitative aspects of the information problem, starting with the importance of different types of publications, and the growth of astronomical literature. Then we shall consider certain aspects of scientific papers, such as their average life, and questions regarding citation. Finally we shall consider some uses of publications for evaluation.

12.1 The relative importance of different types of publications

We have considered in Chapter 11 four major types of astronomical publications: books, journals, observatory publications and minor publications. Let us start by assessing the volume of publications of each type. To do so, we take a simple approach consisting of statistics covering four separate years, namely 1900, 1925, 1950 and 1975; the dates were chosen so as to avoid the two world wars. We shall use as sources the volumes of the *Astronomischer Jahresbericht* (AJ) and *Astronomy and astrophysics abstracts* (AAA) corresponding to these years and I shall introduce some quantities (indicators) that bear upon our problem. I have chosen as indicators:

(a) the total number of pages of the volumes of the AJ and the AAA. The list of observations of asteroids and variable stars, given in the first volumes of AJ but not in the later ones, were excluded;

(b) the number of books published;

(c) the number of astronomical journals published;

(d) the number of meetings held. This is a rather unreliable quantity because, in any year, one gets both the proceedings of the meetings held in the previous year and references to those in the year under consideration;

(e) the number of 'Observatory publication series' existing. We excluded from the count all those publications in which only 'reports of activity' are given;

(f) the number of authors quoted. Statistical corrections were applied when this quantity was not provided directly in the introduction to the volumes of AJ and AAA.

It is clear that all these indicators present some difficulty of their own. Point *a* is affected by changes in style and typography and the figures given are thus not strictly comparable. Points *b* and *c* are uncertain because of the inclusion of border-line journals and books which may or may not have been included. Similarly, point *e* is uncertain because a given observatory might not have published in the particular year chosen. And finally, *f* is uncertain because many authors of papers on astronomy are not themselves astronomers and would not have appeared as co-authors had the paper been published at an earlier time.

In view of these difficulties, it is best to consider the indicators as trend markers, but not to attach too much weight to the individual figures. I remark in passing that I have not tried to introduce any indicator of the 'grey literature', the volume of which is very hard to judge, although it contributes much to the flood of papers.

A glance at the figures in Table 12.1 shows a decline or standstill between 1900 and 1925 due to the effects of World War I, and a sustained growth afterwards, with a duplication period of less than 25 years. Only the number of observatory publications decreases after 1950 for the reasons outlined in Chapter 11. The sharp increase in the number of authors between 1950 and 1975 is a reflexion of the large number of specialists attracted to astronomy by new techniques (radio astronomy, space

Table 12.1. *Growth of publications in the twentieth century*

	Pages	Books	Journals	Observatory publications	Number of authors	Number of meetings
1900	610	68	110	86	(1400)	16
1925	239	61	68	93	(1000)	19
1950	477	132	118	213	2250	45
1975	1083	310	299	85	13 600	146

Unreliable values are given in parentheses.

Table 12.2. *Number of normalized pages*

Year	No.	Year	No.
1910	454	1950	1033
1920	434	1960	2133
1930	505	1970	6156
1940	625	1980	14057

astronomy) and of the fact that papers are frequently co-authored by astronomer and technicians. Also, the number of meetings has increased sharply in the last few decades.[1]

Such rapid growth is confirmed by another study by Abt (1981b), who carried out an investigation of the trend in three general American astronomical journals – the *Astrophysical Journal* (including *Letters* and *Supplements*), the *Astronomical Journal* and the *Publications of the Astronomical Society of the Pacific*. From his discussion, I have taken the data in Table 12.2, which give the total number of normalized pages (i.e. omitting blanks) as a function of time, at ten-year intervals.[2]

These figures show again the very rapid increase in recent years.

Since there is a qualitative agreement between all sources, one can next try to represent the growth by an exponential relationship. For the data of Table 12.1, one finds for books

$$61 \exp (0.0326(t - 1925)), \tag{1}$$

and for journals,

$$68 \exp (0.0296(t - 1925)), \tag{2}$$

whereas, for the data of Table 12.2 one finds

$$1032 \exp (0.087(t - 1950)). \tag{3}$$

In all three cases, the formulae fit the 1900 data poorly. All three results imply duplication times of less than ten years! For earlier periods, one can add the statistics of Houzeau and Lancaster (1880), who compiled the total number of papers by decade, over more than two centuries (see Fig. 12.1). It is impressive to see exponential growth over the whole period, which can be represented by the formula:

$$14 \exp (0.0232(t - 1695)). \tag{4}$$

Exponential growth thus seems to be a widespread phenomenon. It is, however, false to think that it is specific to astronomy, because it broadly characterizes *all* sciences. In Fig. 11.1 we saw how the total number of scientific journals increased exponentially over more than two centuries, and this trend reappears in each science, be it astronomy, physics, etc.

In Table 12.3, for instance, we have the number of 'Physics Abstracts' over the years. It is plain that the two world wars have produced small standstills, but no diminution, with growth reassuming exponentially afterwards. Similar behaviour is described by de Solla Price (1974) for other sciences.

With all these different results, we can assume that exponential growth is a fact of life, with which we have to live. This has some immediate consequences. We can easily calculate with formula (1) the total number

Fig. 12.1. Number of papers published by decades, from Houzeau and Lancaster (1964). The straight line corresponds to $N = 74 \exp [0.0232 (t - 1695)]$ and represents an exponential increase over 170 years.

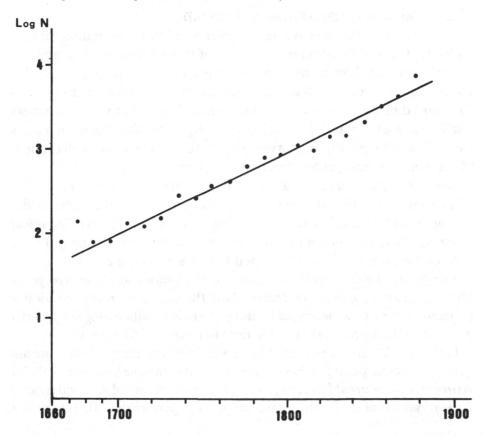

Table 12.3. *Number of 'Physics Abstracts',
according to de Solla Price (1963)*

Year	Log N	Year	Log N
1905	4.11	1930	4.84
1910	4.36	1935	4.97
1915	4.53	1940	5.08
1920	4.61	1945	5.12
1925	4.71	1950	5.21

of books published in the last fifty years, and we obtain 10 647 – let us say, 10 500 books. Since the number of journals is roughly parallel to that of books, we still have to add 10 500 annual volumes of journals, which is a deceptively low figure since (see Table 12.2) the *Astrophysical Journal* alone in 1970 already had 6156 normalized pages, so that the total amount of material to read is simply staggering.

12.2   How much of the information is useful?

It is clear that it is impossible to read all this information and so, logically, the next question is how much of this information is useful? In other words, for how many years is a publication worth reading? This implies that we need a definition of the 'mean lifetime for a publication'. The most direct approach to this is to examine, for a given year, the papers published in a journal and to analyse the frequency with which the papers were cited subsequently. Such a procedure is easy to carry out with the help of the *Science citation index*, described in Chapter 11.

Abt (1981a) has made such an analysis; from his paper, I have taken Fig. 12.2, which gives the 'after-life' for 326 papers published in 1961 in the *Astrophysical Journal* (and its *Supplement*) and in the *Astronomical Journal*. These papers were cited 6070 times in the 18 following years, so that, on average, each paper was cited 19 times, i.e. once per year.

The figure shows a maximum number of citations at about five years after publication, with a slow decline. Half the maximum rate of citation is reached twenty years after publication. This is a rather long life, which contrasts with the much shorter life time in physics of five years.

One may also analyse the citations given in all papers in a given journal over a selected period. I have chosen for the sample one issue of the *Astrophysical Journal* (vol. 288, No. 1) and one volume of *Astronomy and Astrophysics* (vol. 157). The total number of papers analysed is 109, with a total of 2820 references. The results of the analysis are contained in Figs. 12.3–12.7. Fig. 12.3 shows for papers the percentage of citations of papers in each 'age-group' by intervals of five years, and one sees immediately that over 50% of the citations refer to recent (age < 5 years) papers. The 1% level is reached at 15 years. This contrasts with the result in Fig. 12.4, which does the same, but for the books quoted. Here one finds a more even distribution, with a slower decline. This is in line with what we said about the longer average life of a book. Citations of proceedings of meetings are given in Fig. 12.5 and here one finds a very sharp decline with age – the 10% limit is reached at 10 years. This is also in line with what was said in Chapter 11 about the more ephemeral value of proceedings. The results for observatory publications are intermediate, as can be seen from Fig. 12.6.

Fig. 12.7 finally shows how the articles quoted in references are distributed in absolute numbers between publication types – one finds an overwhelming preponderance of reference to other papers.

Fig. 12.2. The annual number of citations to 326 papers published in 1961 in the *Astrophysical Journal* and *Supplements* and the *Astronomical Journal*, shown as a function of years after publication. The standard error of ±19 citations per year is shown as an I in the lower right. The decline after maximum gives a decline rate of −15.4 citations per year or −3.7% of the maximum rate.

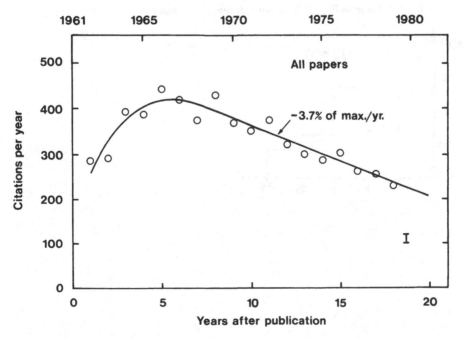

Fig. 12.3. The distribution over time of the references to papers quoted in a paper. Time (t) in years, grouped by five year interval, before publication of the paper. In ordinates, percentage of the total. Total indicated as N on upper right.

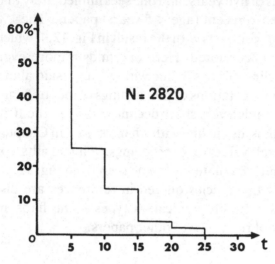

Fig. 12.4. Same as Fig. 12.3 but for references to books.

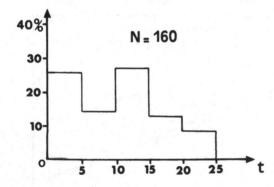

Fig. 12.5. Same as Fig. 12.3 but for proceedings of meetings.

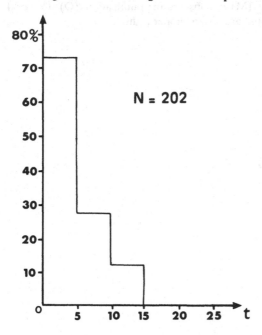

N = 202

Fig. 12.6. Same as Fig. 12.3 but for observatory publications.

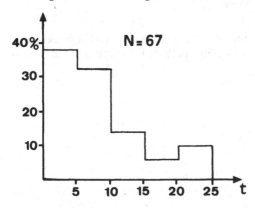

N = 67

Fig. 12.7. Comparative proportions of references quoted in papers to other papers (P), books (B), meetings (M) and observatory publications (O). The total number of references is indicated as N in the upper right.

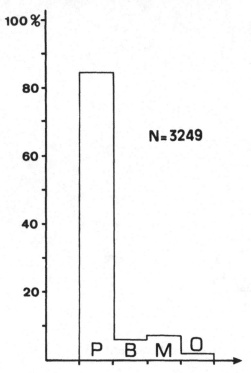

N=3249

Fig. 12.8. Age distribution of the references quoted in a book. The book was printed in 1987 but the manuscript completed in 1985. Time (*t*) is counted in years before 1985.

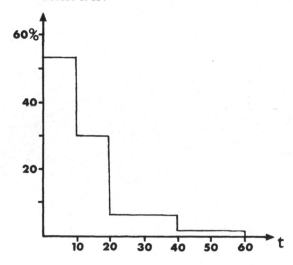

One may deduce similar statistics for the literature quoted in a book. As an example, I have analysed about 300 references to three chapters in a recent book, *The classification of stars* (Jaschek and Jaschek 1987). The percentages are given in Fig. 12.8. One sees again that references to recent work are predominant and that older work is less frequently quoted. The decrease here is rather slow, when compared to the pattern in research papers, so that the 10% limit is reached in 30 years, rather than in 15 years. On the other hand, the most recent references are dated 1985, i.e. two years before the book appeared in print. This is entirely in line with what was said before.

Returning to journals, let us next discuss the apparent contradiction between Abt's result on the average life of a paper and those of Fig. 12.3. A little reflection shows that this is due to the exponential increase of publications with time – our analysis looks, in effect, backwards in time, whereas Abt looks forward. This implies that we are normalizing Fig. 12.3 with respect to a larger number than used by Abt and this reconciles the two considerations. What Abt shows is how many times papers are quoted as a function of their age, whereas our analysis shows, as a function of age, what papers astronomers quote.[3]

The result we demonstrate in Fig. 12.3, namely that more than 50% of references are to papers less than five years old, is interesting in another sense. Merton (1973) obtained similar statistics for other sciences, which are reproduced in Table 12.4.

Obviously, physicists are reading only recent papers, whereas astronomers (53%) fall in line with other natural sciences. In social sciences, the percentage falls to 40% and in humanities to about 20% – obviously the link with history pulls the percentages down.

Table 12.4. *Percentage of papers quoted, less than five years old in different sciences (= PPLFYO)*

Science	Journal	PPLFYO
Physics	*Physical Review*	72%
Biology	*Cold Spring Harbor Symp.*	63
Chemistry	*Analytical Chemistry*	58
Anatomy	*Anatomical Record*	50
Zoology	*American Zoologist*	47

12.3   Relative impact of journals: 'Bradford's law'

Another interesting question that can be asked is whether all astronomical journals are of equal importance or whether an 'order of importance' can be established among them. This is possible with a parameter called the 'factor of impact', which is defined as follows. Consider papers published in years t and t + 1. Find out how many times they are quoted in the year t + 2, and then divide that number by the total number of papers published in the years t and t + 1. The result is the 'factor of impact'. This factor was given by Steinberg (1980) for a number of astronomical journals; the figures are quoted in Table 12.5.[4]

A different study carried out by the author gave a similar result, in the sense that the *Astrophysical Journal, Astronomy and Astrophysics, Monthly Notices of the Royal Astronomical Society*, the *Astronomical Journal, Solar Physics* and *Nature* occupied the first places, if one considers citations only in astronomy and astrophysics in 1985.

Such an ordering depends strongly upon the time and place at which it is made. No doubt, in 1900, the *Astronomische Nachrichten* would have featured at the top of the list (from which it has now disappeared) and the *Astrophysical Journal* would not have been at the top. Moreover, a similar study made in the USSR, for instance, would certainly produce a different order. Another drawback is that such quotations are only applicable to astronomy in a broad sense. If one were to look for specific information in celestial mechanics, for instance, the order would be rather different. The *Astrophysical Journal* would disappear, as would *Solar Physics* and *Nature*,

Table 12.5. *Factor of impact for 1978*

	Steinberg	SCI
Astrophysical Journal	4.35	4.09
Monthly Notices of the RAS	2.71	2.38
Astronomy and Astrophysics	2.31	2.33
Astronomical Journal	2.12	2.20
Icarus	2.03	1.62
Solar Physics	1.83	1.23
Space Science Review	1.55	1.99

From J. L. Steinberg (1980) *Journal des Astronomes Français*, No. 19, p. 2. The second column gives the same data from the 1981 *Science Citation Index* report.

whereas *Celestial Mechanics* would occupy a prominent place. The moral is clear – for each field of speciality (defined by the interests of each astronomer) there are five or six journals that carry most of the essential information, while another twenty or so journals occasionally carry some useful information. Such a situation is again not a specifically astronomical one. Librarians know this problem very well. In a classical paper, Urquhart (1958) analysed the 53 000 consultations of the Central Science Library in London made by outside customers. The Library carried 9120 different scientific journals, of which 7800 are still published. Of these, 4800 were never consulted and only 2274 once in a year. On the other hand, six journals were consulted more than a hundred times. In total, 80% of the consultations could be satisfied with less than 10% of the journals. As can be seen, astronomers participate in this same trend. The general result is that, in any speciality, there are a few 'core' journals, which contain much of the important information, while a larger number of journals provides less information and a very large number of journals provide very little information. This is called '*Bradford's law*'. Roughly, if one says that one third of all information is in the 'core journals', another third is in the 'intermediate journals' and the last third is in journals that contribute only occasional information, the number of journals in each group goes as n, n^2 and n^3.

Should we conclude that most of the literature is useless and that one should only read and publish in the 'top' publications? Although this would certainly help both librarians and astronomers, the answer is clearly *no*. In the first place, it is clear that we do not read only specialized research papers – we all need to read journals with a broad coverage, unless we want to become the superspecialist on HD 124897 who ignores everything outside his own narrow field. So, we need to read summary papers and books, and in a large 'grey' zone in which we find matters of interest to use, but which we will never quote in a research paper. In short, 'top publications' are for our work but not for our general information. Then one must also consider the 'data' aspect, and here the situation is clearly against 'top-publications-only', because observational data appear in *all* kinds of journals; omitting the majority of the journals would mean ignoring the large body of data published in them. With limited time for reading, the best strategy is not to try to read everything (which is impossible), nor to concentrate on reading the 'top' journals, but to use carefully the existing bibliographies, which we have discussed in Chapter 11, from which we may rapidly find out the matters of interest to ourselves.

The second question, whether it is not more profitable for a scientist to publish only in a 'top' journal is harder to answer. One would be inclined to think that, if 'top' journals are more frequently read than others, a publication in one would automatically receive attention. Abt (1981a) has shown, however, in an analysis of two 'top' journals that 6% of the papers contained in his sample were never cited (except in a later publications by their authors). A similar result is reached from an analysis of the *Science Citation Index*, which considers only 'top' publications. Here, almost half of all papers published in the 2100 journals analysed are not quoted at all in the year following their publication (Ziman 1978). So, publication in a 'top' journal does not automatically guarantee more attention to a paper. Furthermore, one must not forget that not all astronomers can publish in 'top' journals, be it because of language, inability to pay page charges or the decisions of national authorities. (Page charges are a system in which the author's institution finances the journal by paying an amount proportional to the length of the paper.)[5]

The best rule to follow seems to be to publish in the periodical of the highest standing one has access to.

We have just mentioned language barriers and we are led automatically to allude to the languages used. Shaw (1985) has given some figures for the languages of publications abstracted in *Physics Abstracts*; for 1972, 68% of the papers were in English and this percentage grew to 90% in 1982. For astronomy, no such data exist but the fact that the most important journals quoted in Table 12.5 publish almost exclusively in English, leaves little doubt that English is now *the* language for research papers. The situation changes when one considers books, observatory reports and 'grey' literature; in these the use of other languages is more frequent. In the Western world, each epoch has had its dominant language – it was Latin up to eighteenth century, French in the eighteenth, German in the nineteenth and will perhaps be an Asiatic language in the twentyfirst century. For written communication, this is certainly an advantage, although for oral communication it constitutes a heavy handicap against those whose mother tongue is not English.

12.4   **Publication and productivity: 'Lotka's law'**

About sixty years ago, Lotka (1926) studied the numbers of papers scientists publish during their lifetimes and concluded that the number of scientists with N papers goes as N^{-2}. This is called the 'Lotka law'. Fig. 12.9 provides an illustration of it, taken from Davoust and Schmadel (1987).

One sees that the law is obeyed rather well, except perhaps for very large N values. The law implies that, if there are n scientists who have published more than ten papers, there will be only $0.1n$ who have published more than a hundred papers; i.e. many scientists contribute a small number of papers, whereas very few contribute many papers. De Solla Price (1963) has shown, on very general grounds, that distinguished scholars tend to be more prolific than others, so that one may state as a rule that prolific writers are distinguished, whereas authors who publish little tend to be more of the average type. One should, however, remember that this is a rule and not a law, because there are very influential astronomers who have published relatively few papers, like W. Baade, and others who have published many papers, like O. Struve.[6]

If one wants to know what 'many' or 'few' means in this context, one can use Lotka's law to calculate the average number of papers an astronomer produces in his life. It is only necessary to fix an upper limit (N_{max}) to the

Fig. 12.9. Number of authors (on logarithmic scale) with n or more papers published over 5 years. Dots for period 1969–73, open circles for 1979–83. Data from Davoust and Schmadel (1987). The line represents the overall fit ($\alpha = 1.7$), which differs from Lotka's exponent.

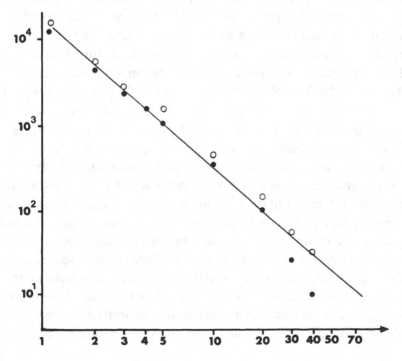

number of papers anyone can write. The average number is rather insensitive to N_{max}; for $N_{max} = 300$, $n = 4.4$; for $N_{max} = 500$, $n = 4.7$. Therefore, the average number of papers a scientist writes is rather small, $n \simeq 5$; this coincides with the average number of publications of scientists in the seventeenth and eighteenth centuries, according to Houzeau and Lancaster (1880).

At present, such an average appears to be very small, because the rate of publication has accelerated so much. Upon closer inspection we find two factors that have changed and are at least partly responsible. The first is the appearance of publications of ephemeral value; the second is multiple authorship.

With regard to ephemeral publications – grey literature, communications at meetings – one must conclude that much material published in this way is later summarized in a regular paper in a scientific journal, so that, typically, one regular paper contains, summarizes and updates results that have appeared previously in part. Due allowance of this fact should be made, for instance, by counting only 'regular' publications in refereed journals. With regard to multiple authorship, it is a fact of life that the average number of authors per paper has increased with the years. Abt (1981b) found, for instance, a steady increase from 1.11 in 1910 to 2.88 in 1980, for observational papers published in three American astronomical publications. For theoretical papers the trend is the same – from 1.0 in 1910 to 1.81 in 1980. This same trend exists in other sciences, where it is even more marked. For instance, Merton (1973) shows that, for physical and biological sciences, the percentage of multi-author papers increased from 25% in the period 1900–1909 to 83% in the 1950–59 period. Except in humanities, the single-author paper is becoming less and less frequent. In multi-author papers it is not easy to find out the contribution made by each author. Usually it is assumed that the order of authors is an indication of their contribution, but it is easy to point out some difficulties. In the first place, scientists often disagree about the importance of their contribution – is, for instance, the 'father of the idea' more important than the one who did the extensive work to examine its worth? Is the author who did the observing more important than the one who did the reductions? Then there are questions of a 'social' nature, such as directors of research units who decide to put their names on any papers to which they have contributed a little, or senior authors who are included because of their past contribution and also technical assistants who contributed to the reductions or observations. Finally, we have papers with six authors (or even share of each person in the paper is essentially impossible to determine. In such cases, authors are usually listed in alphabetical order. This practice seems democratic, but abstracting services or evaluation services usually mention only the first author, in which case the democratic solution becomes an alphabetic solution and first authors obliterate their colleagues. All this shows that, when multiple authorship occurs, it is not easy to decide how a multi-author paper should be counted: as one contribution? as one third (if there are three authors)?

By now the reader will be slightly perplexed as to why such a detailed account should be given of the count of papers. The answer is that 'paper counts' were (and are) used to assess the productivity of a scientist, and therefore to evaluate him or her. To explain this we must make a short digression on evaluation.

12.5   Evaluation of scientific activity

It is clear that the evaluation of scientific activity is important – be it that of a scientist, of a small group of scientists, of an observatory or of a nation. Formerly it was an important problem – today it is crucial. In our times the number of scientists is growing rapidly, but resources tend to have a ceiling; the competition is thus fierce and the hiring or financing institutions are obliged to adopt evaluation techniques to see how they can best allocate the posts and resources available. This means that each scientist should be acquainted with evaluation techniques because they may have a deep influence on his or her own life, or research.[7]

Upon reflection, it seems obvious that criteria will differ according to whether one is looking for *a*) a director of a big institution (for instance, the European Southern Observatory), *b*) a professor of astronomy, *c*) a fellowship candidate, or *d*) an assistant. Since the subject of evaluation cannot be discussed here at length, we shall consider only case *d*), which is probably the simplest. Evaluation usually proceeds by each candidate being asked for a curriculum vitae, which includes his personal data (age, sex, nationality, academic degrees etc.) and an account of his activities (research, position, papers published, honours etc.).

The candidates must first of all satisfy a number of administrative criteria, which may involve age, sex, nationality and academic degrees and which eliminate a considerable number of possible candidates – and there are still health, residence and salary factors, which eliminate other aspirants.

The second part of the curriculum vitae is then analysed for the remaining candidates. The individual accounts of the candidates' research activity are scrutinized but, obviously, these accounts are subjective and difficult to intercompare. A criterion that seems much more objective is consideration and evaluation of the candidate's publications. Criteria based upon publications are called 'bibliometric criteria'. First of all, one can count the number of publications. In consideration of what has been said already, publications in refereed journals are given more weight than others, such as communications to meetings, or preprints. Candidates who have a large number of publications are obviously more productive than others who have fewer, provided one takes into account that younger candidates clearly cannot have produced as much as older ones. This can be done by simply taking the 'average number of publications per year'.[8]

On the other hand, it is clear that numbers of publications do not tell us anything about the relative importance of the publications counted. If a candidate has published an important paper or made a discovery, this should be taken into account. Bibliometrists have thus developed another technique for assessing the 'impact' of a paper. In analogy to the 'factor of impact' of a journal, one may define the 'impact' of a paper as the number of references found in the literature to that paper. If a paper published by Mr X in 1956 is quoted many times in the subsequent literature, it had a large impact. The larger the impact is, the worthier the author is – although, for obvious reasons, self-citations are to be excluded. The impact can be assessed with the help of the *Science Citation Index* (SCI) mentioned in Chapter 11. The drawback of the method is that recently published papers are disadvantaged when compared with older papers, which, if important could have been quoted more times.

Another way to estimate the 'impact' is to see whether one of the candidates has written a paper that has been quoted many times – one can define 'many' as being arbitrarily 50 or 100 times. Such frequently-quoted papers being rare, the rationale is that they are very important.

Although the 'impact approach' is reasonable, there are some limitations. First of all, one has difficulties with multi-author papers. If the author whose impact one wants to assess is not the first author of a paper, one must retrace the 'impact' under the name of the first author of the paper; not doing this heavily biases the statistics.

Then we have the ill-defined difficulty that results from our (unproven) assumption that authors quote past literature exhaustively with strict objectivity – i.e. only the important papers and all the important papers. It is rather dubious if this is so; usually astronomers only know the history of the subject since the time of their own PhD thesis and have marked preferences for the papers written in their own language, by friends, teachers, superiors and colleagues they want to criticize. The work of competitors is often quoted rather summarily or 'forgotten'. It thus remains to be proven that bibliographic references constitute an objective way of assessing the importance of past papers. (See, for instance, McRoberts and McRoberts (1987).)[9]

It is thus clear that bibliometric criteria can help our evaluation but should not be used exclusively. What else can be done? One is left with the 'esteem' approach, based upon the opinion other scientists have of the candidate. This approach is at the basis of different practices, such as:

(a) judgement by senior scientist(s)
(b) judgement by colleagues
(c) judgement by the community of specialists (officer of professional organization, invited lecturer, invitation to write summary papers, to organize meetings)
(d) judgement by the scientific community at large (medals, prizes, honorary degrees, membership of academies).

It requires no great effort of imagination to see the possible difficulties of this approach. Senior scientists may be inclined to prefer their own students to those of their colleagues. Committees may tend to play safe and choose 'normal' candidates with predictable behaviour; committees are notoriously unreliable at selecting very bright candidates. Honours and esteem may be acquired by merits other than the scientific ones, as everybody knows. Presidents of societies for instance have a social role to fulfill (answering letters, organizing meetings, maintaining relations with other bodies) and scientific merit may not be the outstanding criterion on which they are chosen. Prizes and honorary degrees tend to accumulate on a few, according to the Matthew effect ('for unto every one that hath shall be given, and he shall have abundance; but from him that hath not shall be taken away even that which he hath' – Gospel according to St Matthew).

So in the end, people tend to question the 'esteem' approach. Nevertheless, until very recently it was *the* approach used throughout all the history of science. Its advantage is that it works even when bibliometric criteria fail, as in the case of PhD candidates.[10]

In view of the difficulties inherent to each approach, it seems clear that the evaluation of candidates should be based upon a mixture of criteria – perhaps 'impact', 'esteem' and 'number of papers' in this order of importance. Usually, this permits ranking of the best candidates, who can then be invited for personal interviews, which usually produce the final choice. Here again it is clear that some individuals are more able than others to create a good impression with evaluation committees, so it is also doubtful whether interviews produce an entirely 'objective' selection.

Let us remark finally that excessive reliance on bibliometric criteria is producing a worrying result. If a scientist is judged by the number of papers produced, it is logical that he will try to publish as much as he can: 'publish or perish'. The facts that, in past centuries, an average scientist produced less than ten papers in his life (see above) and that some observatories now consider 'two papers per year' a reasonable production rate, imply that the paper mills will be increasingly busy.

The evaluation techniques discussed above can also be extended to the evaluation of groups of scientists, but this falls outside the scope of the book. Some references for further reading are provided in the Notes to this chapter.[11]

Notes on Chapter 12

1. The number of authors (since 1969) has also been studied by Davoust and Schmadel (1987).
2. With regard to Table 12.2 it must be mentioned also that the average length of papers also increases with time (Trimble 1984).
3. The explanation for the discrepancy between these results and those of Abt is confirmed by Peterson (1988).
4. The factor of impact was also studied by Praderie and Davoust in an unpublished paper, but again only for astronomy in general.
5. On language barriers, a number of polemic papers have been published; the emphasis is on the use of the national language – Spanish, French, Russian or Chinese – instead of English. If the use of a national language is imposed – the author has experienced this with Spanish – it automatically diminishes the possible impact of the paper because Spanish, although spoken by many people, is not understood outside the Hispanic world. Similar remarks apply to Chinese, for instance. English in the 1980s has the same role as Latin in the twelfth century.
6. The publication patterns have been changing in recent times in relation to what Lotka's law predicts. With the 'publish or perish' policy there is little chance that the average number of publications will continue to be as low as five for much longer. It probably still holds if one looks only for major publications published in refereed journals and if all grey literature is omitted. It would be worthwhile analysing this in more detail. When only numbers of papers are counted, Davoust and Schmadel (1987) accept as a reasonable number of papers per year per author, one paper. With an active lifetime of 40 years this would greatly surpass the average number of five papers derived in this chapter. For a 15-year interval

they find that, out of 6.6×10^4 authors, only 1.8×10^4 produced more than five papers. If we derived from their data the average number, we find $n \simeq 4$ (over 15 years). With an active life of 35 years this would imply $n \simeq 9$, again larger than our average of 5.

7. Evaluation constitutes a big problem, which requires a book by itself. The purpose of mentioning evaluation here is to show some of the uses (or misuses) of paper counts. A general introduction to evaluation is given in books on the sociology of science: Merton (1973), Ziman (1978) and Ziman (1984). A plea to use selection criteria other than paper counts is made, for instance, by Dawson (1987).

8. Bibliometric criteria are fairly popular now in astronomy, although they were introduced about three decades ago. Probably one of the explanations of its popularity is that counts are easy to carry out and that one gets a numerical, objective answer out of the counts. Difficulties usually appear when one wants to interpret the results! There is an enormous literature on bibliometric criteria – see for instance, the series *Essays of an information scientist* by E. Garfield (Vol. 1–5), ISI Press, Philadelphia.

9. As can be seen, the difficulty of evaluation based upon impact studies lies in the hypothesis that references to the work of others is done impartially. But everybody knows that one's own work is not sufficiently quoted! For studies on citation patterns see Trimble (1985) and Trimble (1986).

   One rather surprising fact unearthed in citation studies (Davoust and Praderie, unpublished) is the following. The authors made a number count of the sources quoted in different journals and concluded that two European journals (*Monthly Notices of the Royal Astronomical Society* and *Astronomy and Astrophysics*) both have the *Astrophysical Journal* as the most quoted journal in their references, quoted twice as much as the next most frequently quoted journal. One can thus conclude that European astronomers recognize fairly well the contributions of their American colleagues. If one examines what is quoted in the *Astrophysical Journal*, one finds in first place the *Astrophysical Journal*, which is quoted eight times more often than the European journals. The importance of research done on both sides of the Atlantic is thus regarded in a very different way, depending on one's geographical standpoint. If one does impact studies, one should thus make due allowance for this fact.

10. The 'esteem' approach was often criticized, rightfully, because it relies upon the impartiality of judgement of scientists; this assumption is sometimes open to doubt. Disturbing tales are heard that having an important godfather can often be more helpful to the career of somebody than other qualifications.

    For scientific prizes, see, for instance, *Nature* **246** (1973) *A medal-seekers guide* (Anonymous).

    For a rather disturbing tale on the working of scientific influence see Dufour (1987).

    In recent years it has become fashionable to criticize esteem approaches and the behavior of scientists in general. When scientific communities grow large and much money is at stake, it is difficult to see why they should not parallel the behaviour of other comparable human groups. Probably, most scientists grew up with hero-worship for great scientists and discovered one day that even great scientists are only people.

11. The evaluation of scientific institutes is discussed in Andrews (1979). A paper that is recommended is Martin and Irvine (1983). The authors analyse in some detail the pitfalls of the different approaches and study one case, namely an inter-comparison of four radio astronomy observatories.

References

Abt, H. A. (1981a) *PASP* **93**, 207

Abt, H. A. (1981b) *PASP* **93**, 269

Andrews, F. M. (ed.) (1979) *Scientific productivity: the effectiveness of research in six countries*, Cambridge University Press

Bradford, S. C. (1950) *Documentation*, Public Affairs Press, Washington DC

Davoust, E. and Schmadel, L. D. (1987) *PASP* **99**, 700

Dawson, N. J. (1987) *Nature* **327**, 550

Dufour, J. P. (1987) *La Recherche* **190**, 940

Garfield, E. (1977) *Essays of an information scientist*, vol. 1–5, ISI Press

Houzeau, J. C. and Lancaster, A. (1880), *Bibliographie générale de l'astronomie jusqu'en 1880*, New edition by Dewhirst, D. W., The Holland Press (1964)

Irvine, J. and Martin, B. R. (1981) *La Recherche* **12**, 1406

Jaschek, C. (1988) to be published

Jaschek, C. and Jaschek, M. (1987) *The classification of stars*, Cambridge University Press

Lotka, A. J. (1926) *Journal of the Washington Academy of Science* **16**, 317

Martin, R. and Irvine, J. (1983) *Research Policy* **12**, 61–90

Merton, R. K. (1973) *The sociology of science*, University of Chicago Press

McRoberts, M. H. and McRoberts, B. R. (1987) *Nature* **327**, 456

Peterson, Ch. J. (1988) *PASP* **100**, 106

Shaw, D. F. (1985) In *Information sources in physics*, Shaw, D. F. (ed.), Butterworth, p. 15

Smith, D. C., Collins, P. M. D., Hickes, D. M. and Wyatt, S. (1986) *Nature* **323**, 681

Solla Price, D. de (1963) *Little science, big science*, Columbia University Press

Solla Price, D. de (1974) *Science since Babylon, enlarged edition*, Yale University Press

Steinberg, J. (1980) *Journal des Astr. Français* **9**, 2

Trimble, V. (1984) *PASP* **96**, 1007

Trimble, V. (1985) *QJRAS* **26**, 40

Trimble, V. (1986) *Czechosl. J. Physics* **B36**, 175

Urquhart, D. J. (1958) *Use of scientific periodicals*, Nat. Acad. of Sciences, Washington

Ziman, J. (1978) *Reliable knowledge: an exploration of the grounds for belief in sciences*, Cambridge University Press

Ziman, J. (1984) *An introduction to science studies*, Cambridge University Press

13

International data organizations

In this chapter we shall consider international bodies that deal with data and data-handling in astronomy. Basically, there are three – the International Astronomical Union (IAU), the Committee on Data for Science and Technology (CODATA) and the International Council for Scientific and Technological Information (ICSTI). I shall briefly describe their objectives and organization, with emphasis on data activities in the case of the IAU and on astronomy in the case of the two others. The reader who wants more details should consult the bibliography listed in the Notes on this chapter, or address himself to the Executive Secretariats of the Organizations.

We shall start with a description of the general organization that coordinates the work of all three.

13.1 The International Council of Scientific Unions (ICSU)

ICSU was created in 1931, to replace the International Research Council. It is a non-governmental scientific organization whose main objective is to encourage international scientific activity for the benefit of mankind, through a network of scientific Unions, Working Groups, Commissions, and interdisciplinary research programmes.

ICSU is governed by a General Assembly, composed of national representatives, the international scientific Unions, scientific and national associates and observers from various bodies.

The General Assembly meets bi-annually and elects the General Committee and the Executive Committee. The latter has its Secretariat at 51, boulevard de Montmorency, 75016 Paris, France.

There are twenty international scientific unions; among them are the International Astronomical Union (IAU) and the International Union of Geodesy and Geophysics (IUGG), as well as many others, such as the

International Union of Radio Sciences (URSI). Among the committees and services one finds, for instance, the Committee on Space Research (COSPAR), the Committee on Data for Science and Technology (CODATA), the Federation of Astronomical and Geophysical Services (FAGS) and also the (Geophysical and Solar) World Data Centers (WDC). Among the scientific associates one finds, for instance, the International Council for Scientific and Technical Information (ICSTI). Seventy-two nations are represented in the General Assembly by national associates.

ICSU publishes an *ICSU year book*, which provides information on its affiliated organizations and a calendar of scientific meetings, and an ICSU newsletter.

We pass next to the Scientific Unions that are concerned with astronomy, namely the International Astronomical Union (IAU) and the International Union of Geodesy and Geophysics (IUGG).

13.2   The International Astronomical Union (IAU)

The IAU was founded in 1920 in the wake of the First World War, synthesising a number of cooperative efforts by astronomers in the nineteenth and early twentieth centuries. According to its statutes, its objectives are

(a) to facilitate relations between astronomers of different countries where the organization of international cooperation is useful or necessary;

(b) to promote the study and developments of astronomy in all its departments.

These objectives are carried out within the general framework of the International Council of Scientific Unions.

Adherence to the IAU is by countries, each country being represented by its National Committee on Astronomy, or an equivalent organization. At present there are 51 national members.

The Union is composed of individual members, admitted on the basis of their achievement in some branch of astronomy; at present it has some six thousand members.

The Union has a number of Commissions – about 40 at the present – each one devoted to the study of a special branch of astronomy, to the encouragement of collective investigations and for the discussion of questions related to international agreements or to standardization. Each

Commission has an organizing committee (usually composed of ten elected members), a President and a Vice-President; both hold office for one term of three years. Commissions may be rather large, as membership goes; Commission No. 40 (Radio astronomy), for instance, has more than 600 members.

The Union organizes General Assemblies every three years, which are held in different countries. These Assemblies are the largest meetings in astronomy and are usually attended by between one and three thousand scientists.

The Union as such is governed by an Executive Committee, composed of nine elected astronomers: a President, six Vice-Presidents, a General Secretary and an Assistant Secretary.

The activities of the IAU can be appreciated through its General Assembly *Proceedings*. These include the preparatory *Reports on Astronomy*, in which each Commission describes the most noteworthy developments in the field during the three preceding years. These reports thus reflect the 'state of the art' and usually constitute a good summary of a research area. After the Assembly, the proceedings of the meetings are published, which also include the 'highlights' (i.e. conferences on subjects of particular interest).

For the rapid circulation of IAU news, the Secretariat edits an *Information Bulletin*, which was founded in 1959 and appears biannually.

Besides the General Assemblies, the IAU also sponsors Symposia and Colloquia proposed by one (or more) Commission. IAU sponsorship of a meeting implies a guarantee of scientific quality and international coverage, and provides a small financial contribution towards its organization. In the case of IAU Symposia, it also implies publication of the proceedings by the official publisher, D.Reidel in the Netherlands. Colloquia do not have such a guarantee of publication. At the time of writing, over one hundred and ten IAU Symposia and over ninety Colloquia have been organized.

Every ten years or so the IAU publishes an *Astronomer's handbook* (last edition, 1966), which provides a description of the IAU (history, finances, administration, interrelations with other scientific Unions, activities and publications), a style book (how to prepare manuscripts for printing, notations, units, etc.) and a list of authorities, members and Commissions.

A list of addresses of members can be obtained from the publisher, D. Reidel. Information on the IAU can be obtained from the Secretariat: 61, boulevard de l'Observatoire, 75014 Paris, France.

A directory of professional astronomical institutes (IDPAI) has been published by Heck and Manfroid (1987). It provides addresses, telephone and telex numbers. The same authors have also published in 1987 a directory of astronomical associations and societies (IDAAS). Both directories are regularly updated, and can be obtained from the CDS, 11, rue de l'Université, 67000 Strasbourg, France.

We turn next to matters of data. There are two Commissions within the IAU dealing with data, namely Commission 5 – 'Documentation and astronomical data' and Commission 14 – 'Atomic and molecular data'. The first deals with data in general, as the title indicates. Because of the breadth of the area covered and the resulting heterogeneity of members – data specialists and librarians – the Commission has four Working Groups (WGs), which are:

1. Astronomical data
2. Designations
3. Classification
4. Abstracting guide lines.

The field of activity of each WG may be summarized as follows. WG 1 deals with data in general, data handling and data centres. WG 2 deals with all matters of object designation; some of its activities were reviewed in Chapter 6. WG 3 deals with the establishment of an IAU vocabulary that can be used for information retrieval through the use of key words. WG 4 provides guidelines for abstracting papers.

Commission 5 has started a Newsletter (*Activities in documentation and astronomical data*), edited by G. A. Wilkins, Royal Greenwich Observatory, UK, and published in the *Bulletin d'Information du Centre de Données Stellaires*.

Commission 5 has also co-sponsored meetings on data centre activities, namely: IAU Colloquium 35, *Compilation, critical evaluation and distribution of stellar data* (1977) edited by C. Jaschek and G. A. Wilkins and IAU Colloquium 64, *Automated data retrieval in astronomy* (1982) edited by C. Jaschek and W. Heintz both published by D. Reidel.

Commission 14, 'Atomic and molecular data', deals exclusively with the kinds of atomic and molecular data that are useful in astronomy. It is really a Commission on data in physics. It publishes extremely useful bibliographies covering a field that is in general not well-known to astronomers.

Three more WGs should be mentioned, which cover other aspects of data. The oldest one is the WG of Commission 45 (Stellar classification), named 'Spectroscopic and photometric data'. As the title indicates, this WG is concerned only with a subset of astronomical data. Probably its most important activity was the publication of a series of lists, *Catalogues recently published, to be published or in preparation*, issued in the *Bulletin d'information du Centre de Données Stellaires* during the years 1977 to 1982.

A second WG is the one on 'Standard Stars'. It is sponsored by Commissions 29, 30 and 45 (i.e. Stellar spectra, Radial velocities and Stellar classification). It covers all matters concerning standardization in stellar astronomy and publishes a newsletter (*Standard Star Newsletter*) edited by L. Pasinetti, Dipartamento di Fisica, Universita di Milano, Via G. Celoria 16, 20133 Milano, Italy.

The most recent WG is the one on 'Modern astronomical methodology', which deals with the fields of statistical methodology, pattern recognition, simulation and other computational problems. It also covers data handling. It produces a *Newsletter*, edited by A. Heck and F. Murtagh, which is published in the *Bulletin d'information du Centre de Données Stellaires*.

When reviewing the data activities of the IAU, one must remember that the IAU cannot impose resolutions, actions or structures. Because of its own structure, the IAU is the expression of the will of the astronomers composing it. At most, the IAU can put its weight behind some actions through its moral and scientific authority, as for instance, when it sponsors meetings or new Commissions.

In view of this, it is perhaps not surprising that the IAU structures are sometimes behind the times. So, for instance, there is a Commission on 'Astronomical Telegrams' (No. 6), but no Commissions concerned with networks, computers, standardization, data archives (or archives of any kind) and with the statistical treatment of large quantities of data. It is sometimes unclear if this is so because of the lack of a sufficiently large number of astronomers interested in such activities, or because of the fact that most astronomers – including those of the IAU Executive Committee – consider such subjects to be of little importance when compared to others.

The new generations of astronomers should be aware of the present shortcomings and try to find remedies.[1]

Before concluding this section, we come back to the problem of the field covered by the IAU. According to its statutes, it covers 'all astronomy', which means in particular that it includes the solar system, which has been excluded from this book. The definition of what astronomy covers is however changing with time. Geodesy, once a branch of astronomy, is now distinct, as is geophysics. If astronomy used to be defined as the science of everything beyond the Earth's surface, it is now the science of everything outside the Earth's envelope and in the near future will exclude the bodies of the solar system, except perhaps the Sun.

Such a nebulous borderline implies that some subjects are studied simultaneously by at least two different scientific Unions. The most important organizations on the fringes of astronomy are

> IUGG – International Union of Geodesy and Geophysics (see below)
> URSI – Union Radio Scientifique Internationale
> IUGS – International Union of Geological Sciences
> COSPAR – Committee on Space Research (see below)
> SCOSTEP – Scientific Committee on Solar–Terrestrial Physics.

All these Unions or Committees have at least one Commission devoted to matters that once belonged to astronomy or are still considered to be astronomy in the broad sense (i.e. including the solar system). In particular, we note that the URSI has a Commission on radio astronomy; this Union also deals with matters concerning frequency allocations for terrestrial users, to avoid interference with extraterrestrial signals. The IUGS has a Commission on 'Comparative planetology'. The SCOSTEP is concerned with all kinds of solar–terrestrial relationships. It also covers the World Data Centers. In a broad sense, it is an Inter-Union Committee that links six unions (IAU, URSI etc.).

The remaining two organizations of the list, namely IUGG and COSPAR will be discussed in more detail later.

A very special relationship exists with another Union, the International Union of Pure and Applied Physics (IUPAP). Since astronomy is, in a wide sense, a part of physics, from which it differs in the choice of the objects of its study and in the methods used, the interrelation between the two is so intimate that a division is artificial. So, we simply mention in passing that it has Commissions on 'Cosmic Rays', 'Astrophysics' and 'general relativity and gravitation', which overlap considerably with astronomy. More specific information can be obtained from its secretary, J. J. Nilsson, Institute of Theoretical Physics, Chalmers University, S – 412 96 Gothenburg, Sweden.

13.3 The International Union of Geodesy and Geophysics (IUGG)

This Union was established in 1919, from an amalgamation of previously existing bodies. The IUGG deals with the study of the Earth and its environment, and the application of this knowledge for the benefit of mankind. This includes the coordination of physical, chemical and mathematical studies of the Earth and its surroundings: the form of the Earth, its gravitational and magnetic field, the Earth's structure, composition and tectonics, the hydrological cycle of water, the study of the ocean, atmosphere, ionosphere and magnetosphere, the Sun–Earth relationship and the Earth–Moon relationship.

The adherence to IUGG is by countries; there are 78 national members. The national delegates and the delegates from the seven associations form the General Assembly, which meets every four years and rules the Union. The administration of the IUGG is in the hands of a Bureau, an Executive Committee and a Finance Committee.

The IUGG is formed by seven semi-autonomous Associations, each one responsible for a certain chapter of geophysics. The complete list is:

- International Association of Geodesy
- International Association of Seismology and Physics of the Earth's Interior
- International Association of Vulcanology and Chemistry of the Earth's Interior
- International Association of Geomagnetism and Aeronomy
- International Association of Meteorology and Atmospheric Physics
- International Association of Hydrological Sciences
- International Association of Physical Sciences of the Ocean.

Each association has its own meetings (often in collaboration with others) and organizes its own research programmes and colloquia.

The IUGG has organized a number of interdisciplinary research programmes, in collaboration with numerous countries, for example the *International Geophysical Year* (1957–58), the *Study of the Earth's Superior Mantle* (1964–70), the *Geodynamics Project* (1972–79), the *Global Atmosphere Program* (1970–80) and the *International Program of the Lithosphere* (1981–90). It has created under the auspices of ICSU a world-wide net of World Data Centres, the data from which are available to all scientists.[2]

The IUGG publishes the *Transactions* of the General Assemblies and a bimestrial *Chronicle* providing latest news. Each of the Associations that belong to the IUGG publishes its own newsletters, news or bulletins. Detailed information can be obtained from the Secretary General, Prof. P. J. Melchior, Observatoire Royal de Belgique, Avenue Circulaire, Bruxelles B–1180, Belgium. Requests for publications should be addressed to the IUGG Publications Office, 39, rue Gay Lussac, 75005 Paris, France.

13.4   The Federation of Astronomical and Geophysical Services (FAGS)

We now pass briefly to an organization that deals with data and was established jointly by the IAU, UGGI and URSI (Union Radio Scientifique Internationale), namely FAGS, Federation of Astronomical and Geophysical Services. FAGS was established officially by ICSU (to which the three Unions are affiliated) in 1956. Its Secretary is Dr R. Wielebinski, Max Planck Institut für Radioastronomie, Auf dem Hügel 69, 5300 Bonn, GFR.

The list of services is as follows (with the founding year in parentheses):

- International Polar Motion Service (1895)
- Bureau International de l'Heure (1911)
- International Gravity Bureau (1953)
- International Center for Earth Tides (1960)
- Permanent Service on Mean Sea Level (1933)
- Permanent Service on Geomagnetic Indices (1932)
- Quarterly Bulletin on Solar Activity (1928)
- Permanent Service on the Fluctuation of Glaciers (1967)
- International Ursigram and World Days Service (1962)
- Centre de Données Stellaires (1985)

Of all these services only the last one, the CDS, falls within the scope of this book. I have mentioned its activities several times throughout this book, specially in Chapter 10, so I refer the reader to that chapter.

13.5   Committee on Data for Science and Technology (CODATA)

It was established by ICSU and UNESCO (United Nations Educational, Scientific and Cultural Organization) in 1966.

The objectives of CODATA are:

- improvement of the quality and accessibility of data, as well as the methods by which data are aquired, managed, and analysed;

- facilitation of international cooperation among those collecting, organizing, and using data;
- promotion of an increased awareness in the scientific and technical community of the importance of these activities.

CODATA has three types of members, namely national members, international members, international Union members and coopted members. Among the 18 national members are practically all the industrialized countries. Among the 16 international Union members one finds the International Astronomical Union and the International Union of Geodesy and Geophysics. Coopted members include the ICSU Panel of World Data Centers, the Federation of Astronomical and Geophysical Services and the International Council of Scientific and Technical Information (ICSTI). Furthermore, there are a number of supporting members, such as Pergamon Press.

The headquarters of CODATA are at 51, boulevard de Montmorency, 75016 Paris, France. Results of CODATA activities appear in the *Bulletin* (obtainable by subscription), the quarterly *CODATA Newsletter* (free), of which more than 36 issues have appeared, and the *Proceedings* volumes from the biennial CODATA conferences, held from 1928 onwards. Proceedings were first published by Pergamon Press and are now done by North Holland; the proceedings of the last conference fill more than 550 pages, which in some way reflects the impact of such meetings.

Besides this, CODATA has also published monographs on various topics in data base management, such as *Data handling for science and technology: Overview and source book*, edited by S. A. Rossmassler and D. G. Watson, published by North Holland in 1980.

Some of the current CODATA projects include a 'Referral data base', which is a computer-searchable file of world-wide sources of data in all fields of science that will include data centres, repositories and other institutions providing data to the technical community. This project is being developed in cooperation with UNESCO. Another important project is the 'Fundamental Physical Constants'. A task force of experts has established the recommended values of the fundamental constants, which are adopted by most national and international bodies.

In the field of astronomy, CODATA has published in its *Bulletin No. 36* a list of astronomical data centres, compiled by Jaschek (1980) and in *Bulletin No. 46* a guide to the presentation of astronomical data, by Wilkins (1982). It has also co-sponsored, with IAU Commission 5, the international course *Data handling in astronomy and astrophysics*, edited by B. Hauck and G. Sedmak and published in the *Mem. Soc. Astron. Italiana*, **56**, Nos. 2–3 (1985).

13.6   Committee for Space Research (COSPAR)

COSPAR was created by ICSU in 1958 to continue the cooperative programmes of rocket and satellite research undertaken during the International Geophysical Years. Its objectives are the progress on an international scale of all kinds of scientific research carried out with space vehicles, rockets and balloons.

The membership of COSPAR is composed of representatives from national Research Councils or Academies and representatives from scientific Unions. There are 36 nations represented and 12 scientific Unions, among them the IAU, IUGG and URSI.

The organization is ruled by the Plenary; its authorities are the Bureau and the Executive Council. The address of the Executive Secretariat is COSPAR Secretariat, 51, boulevard de Montmorency, 75016 Paris, France.

COSPAR organizes Plenary meetings every three years; the *Proceedings* are published by Pergamon Press.

COSPAR has a number of interdisciplinary Commissions and Panels, among them four Commissions on solar-system research and one on 'Research in astrophysics from space'.

13.7   International Council for Scientific and Technical Information (ICSTI)

This Council was established in 1984 as a successor to ICSUAB, the International Council of Scientific Unions Abstracting Board. The purposes of ICSTI are to increase accessibility to, and awareness of, scientific and technical information, and in particular:

- to identify and analyse the requirements of the users of scientific information;
- to study and analyse methods of collecting, storing, organizing and disseminating scientific and technical information;
- to advocate improvements in the sources of, and the systems for providing international access to, scientific and technical information;
- to promote, encourage and undertake activities directed to the achievement of these purposes.

The membership of ICSTI is organized in four classes:

- Class A (Full members) includes organizations representing the interests of information users – scientific Unions, learned societies, national Academies. The IAU and CODATA are members of this type.
- Class B (Full members) are organizations with principal activities in collection, storage organization and dissemination of information – abstracting services, data centres, libraries. *Astronomy and Astrophysics Abstracts* and the Fach-Informations Zentrum fur Energie, Mathematik, Physik (FIZ) are members of this type.
- Class C (Associate members) are organizations with interests in common with those of ICSTI.
- Class D (Honorary members) are persons whose outstanding contribution to the work of ICSTI merits special recognition.

The address of the Executive Secretariat of ICSTI is 51, boulevard de Montmorency, 75016 Paris, France.

ICSU AB and ICSTI have undertaken a wide range of activities, among which we mention a few that have some bearing on astronomy:

- publication of the *International Classification System for Physics*;
- publication of the *International Serials Catalogue*;
- study of the description of data base services through a standard format;
- symposia on various subjects, such as *The online revolution in information* in 1978;
- studies on serial usage and exchanges of experience in the use of new technologies (optical digital discs, satellite communications).

As can be seen, most of these activities concern information, but not specifically data. These are only of marginal interest to ICSTI and correspond to the area covered by CODATA. Recently a 'special interest group on numerical data' was created by ICSTI to further its relationship with CODATA.

13.8   Conclusion

As the reader can appreciate, there are a number of organizations dealing with data. Curiously, most astronomers are reluctant to move outside the IAU, so that organizations like CODATA are left aside by most professional astronomers. This is certainly a very provincial outlook, which was perhaps justified at one time, but is certainly obsolete now.

Astronomers used to think that astronomy is THE science – all written in capitals. I have insisted many times in this book that such a viewpoint is wrong and that astronomers are often not forerunners but simply users of developments in other branches of sciences. Examples can be quoted at length; instrumentation for the detection of X-ray, ultraviolet, infrared and radio sources, computer facilities, data-handling techniques and so on, all came from outside and were merely adapted to astronomy. Because of this, astronomers should look outside their own field much more than they have done up to now, and this also applies to organizations outside the IAU.

Notes on Chapter 13

The best general summary of international scientific organization is the *ICSU Year Book*, which can be obtained from the ICSU Secretariat.

1. The history of the IAU is summarized in the *Astronomers Handbook*, and some personal notes, specially on the first years of the IAU, are given in *Information Bulletins* 22 (1969) and 23 (1970) on the occasion of the 50th anniversary of the foundation. What is still missing is a critical history of the IAU and appraisal of its role, written by an outsider.

2. The ICSU–IUGG World Data Centers are detailed in IAU Colloquium No. 35 (Jaschek and Wilkins 1976).

Index

AAA 150
AAVSO 46
abstracting journals 149
Alexandrian Museum 11
archives 37
  definition 38
  degradation 37
  elements 38
  history 35
  main requirements 47
  observatory archives 44
  plate archives 44
  reduced data 46
  reduction procedure 41
  support media 37
  unpublished observations 158
  what to preserve 40
  why to preserve 40
ASTRONET 140

best value 31
bibliographic inaccessibility 51, 94
bibliographies, astronomical 149
  by object 151
  machine-readable 151
  specialized 151
bibliometric criteria 178
BIIEA 126
books 144, 145, 162
Bradford's law 172

Carte du Ciel 36
catalogues 52, 74
  absolute 52
  after-life 55
  availability 83
  bibliographic compilation 52
  compilation 52
  computer-readable versions 56
  critical compilation 52
  deficiencies of coverage 83, 85

  definition 52, 53
  elements 53
  errors 55
  fundamental 52
  general compilation 52
  header 57
  non-stellar data 80
    cluster data 80
    clusters of galaxies 82
    galaxies 81
    Magellanic Clouds 82
    nebulae and globules 80
    parameters of galaxies 81
    quasars 82
    radio and X-ray sources 82
    special types of galaxies 81
  observational 52
  status code 130
  stellar data 74
    binaries 79
    diameters 79
    magnetic fields 79
    masses 79
    parallaxes 76
    photometry 77
    polarizations 79
    positions and proper motions 75
    radial velocities 76
    rotation 79
    spectrophotometry 78
    spectroscopy 78
    variables 79
CDS = Centre de Données Stellaires 126
charge-coupled devices 37
clay tablets 10, 35, 37
CODATA 183, 190
completeness of data knowledge 104
computers 111
Córdoba Observatory 13
correspondence 146
cross-identification 133
COSPAR 184, 192
critical evaluation of data 115

CSI = catalogue of identifications   133 (*see also* cross-identification)

data   24
  acquisition rate   40
  archives   38
  analysed = reduced   28
  best values   115
  calibrated = edited   28
  compilation   156
  critical   32
  definition   24, 27
  edited = calibrated   28
  errors   29
  examples   24
  in teaching   27
  journal   157
  kinds of   27
  presentation   29
  raw data   27
  reduced = analysed   28
  units   29
data bank   110, 112
data base   110
  compatibility   122
  definition   112
  directory   123
  integrity   119
  journalization   120
  presentation   121
  proof reading   119
  resilience   120
  security   118
  standards   121
data centres   126
  access to information   138
  announcement of availability   130
  characteristics   127
  definition   127
  economic aspects   130
  future   141
  legal aspects   132
  manpower problems   131
  standardization   128
data collection   110, 112
  critical evaluation   115
  definition   112
  establishment   112, 115
  updating   114
  storage forms   110
data growth   93
  annual rate   93, 100
  doubling time   93, 102
  compound indicator   104
  exponential rate   93, 100, 164
  for specific data   94
    asteroids   95
    photometric data   97
    pulsars   98

  quasars   98
  spectroscopic binary orbits   95
  trigonometric parallaxes   96
  variable stars   98
  visual binaries   96
  for classes of objects   102
  for types of objects   104
  index of redundancy of certain data   94
  logistic curve   100
  rate   93
data, international organizations   183
data, knowledge completeness   104
  for galaxies   106
  for stars   104
data presentation   29
designation of astronomical objects   58
  and position   62
  and priority right   69
  dictionary of synonyms   62
  dictionary of nomenclature   61
  Durchmusterungen   59
  IAU resolution   65
  IAU rules   64
  official practices   65
  of objects   61
  of stars   58
  of non-stellar objects   63
  rules   61
directory, astronomical   186
  data bases   123
  data centres   133

electronic journal   156
errors, in general   29
  accidental = internal   30
  external = systematic   30
  internal = accidental   30
  magnitude of   31
  systematic = external   30
esteem approach   179
evaluation   177

FAGS   184, 190

grey literature   155
growth of data   93
growth of scientific information   162

header   57

IAU   183, 184
  Commission 5   186
  Commission 14   186
  Working Groups   186
ICSTI   183, 192
ICSU   183

ICSUAB 193
impact 178
importance of journals, relative 172
intellectual property 132
internal reports 153
international data organizations 183
interview 179
invited review paper 154
IUGG 183, 188, 189
IUGS 188
IUPAP 188

journal, scientific 148, 162
   relative importance 172

language barriers 174
list 50
literature search 158
logistic curve = sigmoid 100
Lotka's law 174
log book = observing file 36, 39, 43

magnetic tape 37
magnetic disc 37
Matthew effect 179
meetings 154
   colloquium 154
   conference 154
   symposium 154
   workshop 154
microfiche 156
microfilm 38
microprint 155
minor publications 153, 162
models 26
much-quoted paper 178
multiple authorship 176

network 138
   communication facility 138
   public 139
   specialized 140
newsletters 153

observations 1
   coordination 8
   definition 1
   elements of definition 2
   examples 1
   planning 5, 7
   programme 6
   programme type, monographic 6
   programme type, survey 6
observatories 2
   American 14

archives 44
   definition 10, 13, 15, 22
   Greek 10
   history 10
   maximum efficiency 18
   medieval 12
   mega-observatory 16
   modern 19
   mountain stations 14
   multinational facilities 15
   Muslim 11
   national facilities 15
   publications 152, 162
   radio astronomy observatories 19
   space institute 20
   size 17
observing file = log book 36, 39, 43
optical disc 38
observational archives 35

paper 37, 144
paper counts 177
papyrus sheets 37
parchment 37
Paris Observatory 12
PASCAL 113
peer-review system 147
periodical 148
photoelectric photometry 36, 41
plate archive 42
population variance 30
photographic plates 36
presentation of astronomical data 50
primary literature 155
priority rights 69
probable error of the mean 30
proceedings, of learned societies 146
   of meetings 153
productivity 176
program committee 6, 16
prints 146
publication of scientific information 144
   form 155
   and productivity 174
   mean life 166
   usefulness 166
   publish or perish 180

quotations 178

radio astronomy 19
rapid journals 152
reduced data archives 46
reduction procedure and archiving 42
referee 147
reference data journal 157
rejection of discordant data 31

remote operation   21
review paper   154
Royal Greenwich Observatory   12, 17

sample mean   30
    standard deviation   30
satellites, artificial   42
scientific information, publication of   144
scientific information, growth of   162
SCI   151
SCOSTEP   188
sigmoid = logistic curve   100
SIMBAD   114, 134
space astronomy   19
    experiments   45
    observations   45
standard datum   32
standard deviation of the mean   30
standard   32
    fundamental = primary   33

general   32
primary = fundamental   33
secondary   33
STARLINK   140
support media   37
synoptics   158

theories   26
top journal   174
trailer   57
Tycho Brahe   36

units   29
Uraniborg   12
URSI   184

WDC   184